CRACKS IN THE ALLIANCE

Dedicated to my family.

Cracks in the Alliance

Science, Technology, and the Evolution of U.S.-Japan Relations

ANTHONY DIFILIPPO
Professor of Sociology
Lincoln University
Pennsylvania

Routledge
Taylor & Francis Group
LONDON AND NEW YORK

First published 1997 by Ashgate Publishing

Reissued 2018 by Routledge
2 Park Square, Milton Park, Abingdon, Oxon, OX14 4RN
605 Third Avenue, New York, NY 10017

Routledge is an imprint of the Taylor & Francis Group, an informa business

A Library of Congress record exists under LC control number: 97070885

ISBN 13: 978-1-138-61239-6 (hbk)
ISBN 13: 978-0-429-46346-4 (ebk)

Contents

Figures and tables

Preface

This book is an analysis of the relationship between the United and Japan, mainly since the end of World War II. It investigates many of the major issues and problems that have affected this relationship, most notably those related to science, technology, trade and defense. It further proposes a comprehensive agenda for Japan, specifically that it fashion a new approach for a strengthened bilateral relationship in the changed global system. This book argues forcefully that Japan must firmly and unequivocally establish itself as a genuine post-Cold War superpower by actively promoting authentic international security through the United Nations, devoid of U.S. hegemonic influence.

Following the introduction, which briefly examines this book's content and purpose, the early chapters focus largely on the divergent science and technology paths taken by Japan and the United States and show how this has related to ongoing bilateral trade problems. Although the last two chapters change the direction of the book, they are not meant to be analytically separate from its earlier empirical arguments. Rather, the purpose of the last two chapters is to detail the changes that have taken place in the bilateral relationship and in the global system, as well as to introduce the prospects for new directions in this relationship and in the international arena.

I have been fortunate in many ways during the course of my work on this book. A number of people helped me -- indeed, too many to list here -- by providing crucial information. Among them at the Japanese Embassy in Washington, D.C. were Akira Furukawa, Katsuyuki Ohyama, Kimikazu Iwase, Hideo Yamazaki, and A. Fujita. Ken-ichi Iwasaki, Director of the Japan Information Center of Science and Technology in Washington D.C., kindly gave his time and supporting information. In Tokyo, Keiko Horie Chief of the International Cooperation Unit in the Statistics Bureau of the Management and Coordination Agency provided me with historical data on Japanese research and development expenditures and activities. Staff members at the National Science Foundation were also helpful in supplying information for this work, among them Lawrence Rausch who made available to me U.S.-Japan data on high technology trade. The staff at the Office of Technology Assessment was helpful by expeditiously making reports available to me. Working ideas drafted and presented at several professional forums gave me academic feedback on some of my ideas, ultimately enabling me to further develop my analyses. My colleague and friend Chronis Polychroniou was instrumental in arranging a stay for me during the Spring 1994 semester as a Research Associate at Tufts University's Fletcher School of Law and Diplomacy. This appointment relieved me of my teaching responsibilities and gave me the free time necessary to work on this book. While there, I made particularly good use of the library resources of the Documentation Center on Contemporary Japan at Harvard University and, to a lesser extent, its Kennedy School Library. During the course of this work, I also received several small research stipends from Lincoln University. These stipends enabled me to travel to libraries -- including specialized ones, such as the Japan Economic Institute of America and the Sasakawa Peace Foundation USA, both in Washington, D.C. -- to perform some of the research for this book. Certainly not the least of all was the invaluable assistance I received from my wife Patty who exhaustively read the manuscript.

Finally, I must point out that in the 'Notes' section that follows each of the chapters in this book Japanese names are written as they appear in the original text material.

Anthony DiFilippo
December 1996

1 Introduction: the undoing of paternalism

Different from four decades ago, today the alliance between the United States and Japan is in relatively poor shape. Despite the argument which maintains that the diplomatic framework in which these two nations have long operated has produced the impression of ongoing conflict and tension, while at a more fundamental level, where it really counts, relations have been good,[1] the past decade in particular has given rise to a steady stream of serious problems in the alliance. Indeed, the troubled alliance between Japan and the United States has supplanted the tensions that emanated from superpower discord that existed during the Cold War. U.S.-Japan discord stems not from the conflict between two nations that have antithetical economic and political orientations. Rather, it is a intense inter-capitalist rivalry manifested mainly in bilateral trade that decisively demonstrates the importance of economic issues.

Contemporary trade issues, however, relate intrinsically to the possession of science and technology (S&T). Together, S&T has historically been a major investment catalyst of growth, and in today's global economy indispensable to a dynamic industrial nation since it is a primary source of innovation.[2] President Clinton recognized much of this when he recently stated that: 'Technological innovation is the engine driving the new global economy.'[3] With regard to U.S.-Japan relations, it is not just the problems connected to the visible possession of and access to S&T that has strained the alliance, but also the fact that the trade dispute increasingly became the outcome of its application.

In this work it will be shown that this new superpower discord is rooted in the past, and reflects to some extent the paternalism that was part of the Japan-U.S. relationship for decades. This work will also show that the discord that is currently recognizable between Japan and the United States today was inevitable, since their post-war historical paths frequently turned different corners. Japan adopted a peace constitution; the United States was determined to remain the world's leading military power. Japan gave priority to strong and sustained economic growth; the United States had foreign policy interests that were a constant distraction to the development of domestic policies that supported sustained growth. With its strong industrial economy Japan wanted to build a first-rate technology infrastructure, which in time it began to integrate with science; the United States' foreign policy objectives placed it on a much different technological and scientific path. Each problem involving the United States and Japan -- those pertaining to trade disputes affecting a number of industries over the years, to technology and more recently science, and even to international security matters -- have therefore not been isolated incidents which can be easily solved by policy makers in both nations without the willingness to implement significant change.

Although getting off to a very different start from other bilateral relationships (the defeat of Japan through the use of nuclear weapons by the United States and the subsequent occupation, which lasted nearly seven years), the alliance worked for some time. Despite the strong opposition from some in Japan, most notably many members of its socialist party, the alliance worked because of the hysteria connected to the Cold War and the very big differences in the economic strength and technological capabilities of the two nations. These two conditions made it possible for both nations to pursue their own interests for a few decades: those of the United States centering on military and economic hegemony, while those of Japan focusing on building a strong and technologically advanced industrial economy. Though by no means occurring at the same time, nonetheless changes in the two conditions that made the alliance operational precipitated growing discord, which effectively complicated the evident cultural dissimilarities between the United States and Japan.

Some collisions between the United States and Japan did occur from time to time in the past. However, they have become more commonplace in recent years as the tensions of the Cold War subsided and because Japan has become an economic and S&T superpower. Today, the frequent fire storms that we observe are the consequences of both nations moving directly and relatively independently at each other with sufficient momentum to create ongoing and increasingly serious problems in bilateral relations and the international political economy.

The past catches up with the present

Through most of the long relationship between the United States and Japan, there has been one undeniable constant: the mutual security treaty signed in 1951, the core of the alliance. Although the treaty effectively assured Japan that the United States would provide protection if needed, the Japanese reluctantly accepted the arrangement because political and ideological divisions in the United Nations (U.N.) prevented it from assuming this responsibility. Because of this division with the U.N. and the Japanese commitment to pacifism, Japan had only two major alternatives. It could, as it did, accept the security treaty with the United States, or it could adopt a policy of 'unarmed neutrality' supported by many socialists.

Even though the socialist policy was never adopted, many Japanese certainly demonstrated a dislike for America's military protection. The security treaty and the accompanying presence of U.S. bases and troops in Japan were never fully welcomed by the Japanese people. Public opinion surveys during the 1950s showed substantial and generally increasing disapproval of American bases in Japan. Public disapproval reached a high point in October 1957 when 60 percent of the Japanese surveyed expressed opposition to the presence of U.S. bases in Japan.[4] When the security treaty was renewed (and also revised and renamed) in 1960 it erupted into a major political problem and the source of public disturbances.

The steadfastness of the American-inspired position that Japan required protection from the United States became a necessity the Japanese eventually came to accept, giving the alliance a sense of purpose. Another important matter played a part in what was early on a relatively good bilateral relationship. For at least a quarter of a century after World War II, America's economic and S&T dominance had a humbling effect on Japan. These realities together, put Japan in a not particularly favorable position. In a relationship that has long been said to be a partnership, from the American side, Japan rarely has been treated as an equal. Nonetheless, for a number of years the differences that emerged between Japan and the United States were more or less resolved amicably, though with the use of American pressure.

As Japan's economy grew stronger and as its S&T base became increasingly more advanced during the 1970s discord became more salient in the bilateral relation. When Japan acquired the position of an economic and S&T superpower in the 1980s tension grew even thicker. Not long after this the moorings that settled the alliance were becoming weaker as the purpose of the security treaty diminished in importance. Although the security agreement between the United States and Japan remains intact, today it has lost most, if not all, of its meaning. It is, indeed, a relic of the Cold War; a symbol of cooperation in an alliance that has fundamentally changed over the years. Thus by the early 1990s change had produced the effects responsible for the more troubling and obvious encounters in the bilateral relationship. Accompanying this problem has been incessant pressure by the United States on Japan to make changes in the operation of its domestic markets and to adjust its defense effort to American standards.

To some extent these are the effects produced by the United States diving shoulder deep into the Cold War with the former Soviet Union. This meant that a major objective of U.S. foreign policy, which because of the big military budgets simultaneously worked as a domestic policy tool,[5] at the very least precluded serious consideration being given to an array of domestic issues, including an ongoing government-supported commercial S&T agenda.[6] So that critics will not continue to jump on the political bandwagon maintaining that there is no guarantee that in the absence of Cold War involvement U.S. policy would have addressed civilian-industrial concerns, it needs to be stressed that this argument entirely misses the point. The fact that the United States did become absorbed by the Cold War, creating what many have called a *de facto* military industrial policy, is the only important issue since this action eliminated the possibility of implementing a structured and inclusive civilian-industrial policy. Thus the argument here relies on an empirical point, not a hypothetical one as does that of the critics and, just as significantly, a relative condition: what the United States *did* relative to Japan. Moreover, it would be perfectly legitimate to consider the many social and economic problems also left unattended because of the type of policy employed by the United States.[7] What America's policy choice did do is force the nation into a domestic posture that made it politically easy to avoid commercial interventionism. America's chief foreign (military) policy concern also helped to push up the budget deficit. At the same time, structural unemployment steadily became a bigger problem, thereby shrinking the tax base.

Japan's domestic policies, on the other hand, mainly focused on the promotion of economic growth. As its economy surged and its industries matured and grew stronger, its engineers learned how to hone and advance imported technology and its policy makers came to realize the importance of giving the nation's scientists the opportunities to delve into advanced research. Thus for Japan, economic growth became the correlate -- rather than the consequence -- of paternalism. Through a period of extensive growth, Japan remained committed to 'anthropocentricism' or 'peoplism,' as well as to 'employee sovereignty,' permitting it to approach egalitarianism,[8] a frequently unmentioned economic and cultural 'difference' that markedly set Japan off from the United States. Japan's foreign policy also focused on supporting its economic objectives. Content with the constraints placed on it by the peace constitution, Japan adjusted its foreign policy to support the United States through the leadership of the Liberal Democratic Party. Even when uncomfortable with U.S. policy leadership, Japan did not take independent initiative in the formation of Cold War policy such as in issues dealing with the arms race (*gunbi kyoso*) and disarmament (*gunbi teppai*).

Beginning approximately in the mid 1980s revisionist thinkers in both the United States and Japan mainly began to reassess Japan-U.S. relations. Departing from the conventional explanation which emphasized Japan's eventual 'coming around' to Western ways, revisionists in the United States began to stress that the economic differences in Japan were irreconcilable using conventional bilateral mechanisms. Meanwhile, this thinking led to a critical response from Japan. Some Japanese began to rethink U.S.-Japan relations. This new or revisionist thinking in Japan emphasized that America's weakened industrial competitiveness was the direct result of its own shortcomings and that Japan should pursue a more autonomous direction for itself.[9]

Understand that revisionist thinking in both Japan and the United States first appeared in the context of an emerging diminution of Cold War tensions. Importantly, revisionist thinking is a direct result of a major structural shift that began to unfold slowly during the 1980s: because the line between domestic and foreign policy was becoming increasingly blurred as inter-capitalist rivalries were growing more intense, the meaning of national security was undergoing fundamental change, most especially for the United States. Thus 'containment politics' was soon to acquire a new meaning, shifting from the Soviet Union to Japan.

The end of the Cold War initially put U.S. policy makers in an awkward position. It became obvious that new foreign and domestic policies were necessary. The expenditure drain given to the Cold War needed to be reconsidered; there was indeed a real potential for the realization of a genuine 'peace dividend,' and so it became evident that it is was necessary to give more attention to domestic issues. But American policy makers were not really willing to make significant adjustments, and largely became content with the political vacuity of the 'new world order.' Despite the appeal of the slogan, U.S. policy makers for the most part have thus far demonstrated their reluctance to let go of past policy that underscores military superiority. Security adjustments made by the United States have been nowhere near sufficient. It would be far too hasty to be optimistic about the suggestion given by American policy makers during the past several years that meaningful conversion initiatives (from military to civilian economy) are in the making.[10] Similarly, one should not be too optimistic about the contention of a self-initiated policy adjustment by the United States to the post-Cold War environment that shifts policing responsibilities to the U.N., but that requires American leadership.[11]

The reluctance to change on the part of the United States is matched by tentativeness on the part of Japan. The post-Cold War period has increased worldwide expectations that Japan assume more global responsibility. Indeed, the criticism has been frequently and correctly made that Japan is not doing enough in the way of taking on more global responsibilities.

Bilateral problems in political economy: what about before?

It is true that big cultural differences between Japan and the United States have produced some market problems.[12] These cultural problems, however, are not primarily responsible for the changes which have adversely affected bilateral relations. Today the most politicized part of the bilateral discord stems from the position taken by most U.S. policy makers and others that Japanese domestic markets are closed to American products.

While this and related matters will be addressed in detail in Chapter 4, briefly stated, the evidence does not support this position. A simple measurement seriously questions the closed-market hypothesis. In 1969 Japan imported 9.2 percent of America's total merchandise exports. This is somewhat *lower* than the proportion of U.S. total merchandise exports that it accounted for in 1993. It is true that part of the recent growth of American merchandise exports to Japan is the result of Japanese businesses operating in the United States. This, however, does not account for all of the increase in U.S. exports to Japan, particularly when looking at manufactured exports, including high technology products. Thus the question is how was the United States able to be the dominant economic power earlier in the post-war period with a relatively smaller part of the Japanese market then than it acquired in more recent years?

The answer to this question rests on the fact that Japan has become a progressively important player in international markets. It not only has turned aggressively to Asian markets, but it has also increased its export share in Europe. As a proportion of all Japanese exports, those going to Europe have increased from 19.8 percent in 1986 to 21.8 percent in 1992. Using monthly averages, the proportion of Japanese exports going to the European Economic Community have increased from 13.6 percent in 1983, to 18.9 percent in 1988, and to over 20 percent in 1992.[13]

Although subsequent discussions in this work will not address in detail Japan's growing interest in the European export market, it does need to be mentioned that this further questions the credibility of American claims of Japanese unfair business practices. To some extent, increasing Japanese exports to the European market have been 'crowding out' U.S. producers. The bottom line is that many American policy makers have implicitly felt that the military security Japan received from the United States consistently obligates it to purchase more of America's products. Arguably, this includes help in compensating for weaknesses in other parts of the world.

During the 1960s the United States was indisputably the leading economic power in the world. Japan's economic power, while growing rapidly, was then not much of a threat to the United States. The American position was that Japan was a strong developed nation and that, even though there was definite evidence of trade discrimination being practiced by Western nations (but not the United States), it was time for the Japanese to stop protecting their domestic markets.

Japan, on the other hand, considered its economy to be only 'semi-developed' and that while growth was plainly evident its industries were still in infant stages. For Japan, this meant that some protection was still necessary. Policy makers in Japan considered the existing barriers in Western Europe and the United States as a big part of the reason why they had been slow to lift domestic import restrictions.[14] Japan's policy makers eventually understood that increasing imports would not be terribly problematic if these could be offset by growing exports.

Over the years U.S. pressure on Japan to open up its domestic market has been subject to change in both amount and in orientation. In the 1960s American pressure on Japan to open up what were then certainly protected markets was nowhere near as intense as it became in more recent years. The pressure that did exist in the past took the form of attempting to remove formal import restrictions such as exchange license controls; there was no discussion of major impediments to foreign products created by the *keiretsu* (industry groups), or other invisible structural barriers.[15] Today, just how open, or for that matter closed, is the Japanese market really is not the most important question. Rather, the point is that the United States is applying an inordinate amount pressure on Japan to adjust to its standards.

As Japan continued to grow more economically powerful, American policy makers and analysts observed that Japanese imports from the United States were not as high as those of Canada, or those of the combined total for the European nations. According to the dominant American perspective, Japan's imports, including those from the United States, should have increased proportionate to its economic growth. Because they did not, therefore, its markets were closed.

However, this is an 'economistic' assumption, focusing entirely on what Japan's imports *should be* relative to other industrial nations. This disciplined-based provincialism eschews some important facts. Japan has reached a higher level of manufacturing, including high technology, competency than Canada and the European nations. That the Japanese people have a predilection for Japanese products should therefore come as no surprise. Moreover, as a publisher of a U.S.-produced journal on Japan astutely put it: 'Alas, why would Japan's consumers buy what can't be sold at home [in America]!'[16] Quality applications, which were brought to Japan more than four decades ago by William Deming and subsequently evolved to more advanced levels, together with a commitment to national economic objectives (that have yielded an improving lifestyle for the majority in Japan), a commitment which is fundamentally no different from a 'buy American' view, provide an alternative explanation that further weakens the closed-market argument. Add to this the frequent Japanese criticism that American businesses commonly lack the commitment to make adjustments which will bring success in Japan (e.g., automobiles), and the closed-market argument becomes even less credible.

Still more

Problems between the Japan and the United States are not limited to the degree of openness in the Japanese domestic market. Over the years there have been continuous complaints that Japanese manufacturers have 'dumped' products in the United States. This has led to the United States imposing a number of anti-dumping penalties on Japanese products. In recent years the Japanese government has made a point of emphasizing the willingness of the United States to use punitive measures such as anti-dumping penalties, and relative to other industrial nations, to maintain them for a much greater time. The current objective of Japan is to emphasize that America's allegations of unfairness constantly leveled against it are unusually harsh and are more reflective of a protectionist strain in America's managed trade policy than of significant infractions on the part of Japanese manufacturers and government. While not capturing the same attention as the alleged closed-market problem, it is nevertheless evident that Japan has become more willing in recent years to voice its disapproval of America's restrictions on Japanese imports to the United States, including its disdain for the unilateral policy instrument Super 301.

In recent years discussions on managed trade have become an integral feature of bilateral relations between Japan and the United States. Republican administrations in the United States during the 1980s formally endorsed a commitment to free trade. However, what they practiced was something very different -- trade with Japan was definitely managed. One could easily argue that the Clinton administration's Japanese trade policy has even more ostensively embraced a managed trade approach than its predecessors.

It is imperative to stress that while managed trade between Japan and the United States has received much attention in recent years it is certainly not new. The United States negotiated the first 'voluntary' export restraint on cotton blouses manufactured by Japanese producers in 1955. By 1961 a total of 25 Japanese products were subjected to voluntary restraints; this meant that voluntary restraints existed on somewhere between 25 to 35 percent of Japanese exports coming to the United States.[17] While the criticism has frequently been heard that Japan has received a 'free ride' from the United States (by getting military protection at low cost), one could easily argue that the existence of voluntary export restraints on Japanese products (e.g., on automobiles) gave American manufacturers a free ride by helping them avoid even bigger economic catastrophes than they experienced over the years.

Science and technology surfaces as a problem

During the 1980s problems involving science-based technologies became a growing wrinkle in the bilateral relationship.[18] Japan's continued advancement in science-based industrial technology development put the

United States in a politically precarious position. Because America's competitiveness had weakened, reflected largely by the erosion of many markets it had formerly dominated, most observers understood during the 1980s that the United States needed to improve its S&T base, some even recognizing the existing developmental and application differential relative to Japan.

In the meantime a critical point in technology development had been reached: analysts declared that in a growing number of areas commercial technologies had become more advanced than those of the military. While many observers made this point, none sufficiently explained how this occurred since American industry did not demonstrate a continued resolve to make capital investments in S&T, and the U.S. government's interests remained committed mainly to its military objectives. How then did this happen? What should have caused analysts to scratch their heads, instead became widely accepted as a smooth transition in technology development.

In this work we shall see that during the early 1980s the United States began applying pressure on Japan (in addition to that relating to trade-related issues) to acquire its technology for military purposes. As will be shown in Chapter 5, the eventual concessions made by Japan were not sufficient, since later in the decade the United States began demanding 'symmetrical access' to Japanese technologies. This made the whole issue even more complicated than it had been earlier in the 1980s. Having established itself as an economic and S&T superpower, Japan became protective of science-based industrial technologies because many were the property of business.

At the same time there was a reluctance in America to share advanced military technologies with Japan, as in the FSX fighter episode. Some policy makers thought that to do so would jeopardize U.S. industry. Intertwined with all of this was the charge that Japan had acquired a second 'free ride' from the United States because it was not performing enough basic research and so relied too heavily on American investment.

Wither ideology: the standoff

U.S.-Japan relations had been going through a process of quickening deterioration during the 1980s. This occurred for two important reasons. First, Japan's economic challenge to the United States had become very evident. It had expanded its sphere of economic influence. A few developments indicated this. Its growing foreign investments, including (and most problematic for the stability of the alliance) those in the United States. Its export-driven economy had moved beyond traditional manufacturing and was at this time continuing to chip away at international high technology markets. Japan had, in other words, established itself as a formidable, but relatively 'unarmed' superpower. Second, in the latter half of the 1980s U.S.-Soviet relations had improved enough to move the spotlight away from ideological discord and in the direction of a growing inter-capitalist rift.

By 1990 Japan-U.S. relations had worsened, now clearly reflecting the tensions of competing superpowers. Japan had become the major threat to America's national security -- a term that in the United States had gone through a change during the preceding decade, conveying considerably more of an economic meaning than it did in the past. Thus in the minds of some Americans Japan had become a greater threat to their well being than the Soviet Union. At about this same time, this national sentiment in the United States was being matched in Japan, indicated in *kenbei* (dislike America). President Clinton became the main target of Japanese *kenbei*. A June 1993 survey conducted by the Nippon Research Center indicated that 51 percent of the Japanese respondents said that they did not like president Clinton. This was an increase of 17 percent from a poll taken only a few months earlier. This survey of the Japanese public gave Clinton the distinction of being the most unpopular American president ever recorded in Japan.[19]

Japan attained the position of an economic and S&T superpower during a time when global politics were undergoing unprecedented post-war change. It is in this context that Japanese objections to U.S. dominance of the alliance have become increasingly evident. It is also evident that for Japan the price of security -- which technically has been nothing less than compliance to the United States and reaching well into the area of political economy -- has become too high in what is an ideologically neutral period. Thus recently Japan has begun to react assertively to continued U.S. claims of trade unfairness and it has not been willing to provide unimpeded access to the private sector's research work. Moreover, despite the continued pressure from the United States to take on more of the military burden, the growth of defense spending has slowed precisely at the time when Japan is searching for a new international position.

It is important to understand that the argument here is not that Japan and the United States cannot find a common basis for cooperation. There are certainly a number of important examples where cooperation exists. It can be found between the governments of both nations,[20] including in international S&T programs,[21] it can especially be found between Japanese and American corporations,[22] and both nations in principle share the belief for the need to disseminate democracy and human rights, part of the broader meaning of global security. What is being said is that discord is the predominant characteristic of the post-Cold War relationship, precisely because the United States continues to try to dominate it during a time when domination will not work. Even though the international system has gone through major changes during the past several years, the bilateral relationship remains an unequal alliance, with unrelenting foreign pressure, or *gaiatsu*, which is largely reducible to American pressure on Japan, still very much observable. The big difference now is that Japan is much less willing to comply readily as it did in the past.

Developing a new international position

The Japanese have understood for some time that their nation needs to develop an international posture that befits its global superpower status. After taking a tentative step in this direction a few years ago, many Japanese policy makers felt that the U.S. criticism of 'dollar diplomacy' concerning Japan's contribution to the Gulf War was excessively harsh, given the nation's resolve not to engage in international armed conflict. This criticism exacerbated the existing image of Japan as a free rider, and at the time set off an internal debate about its contribution to global security. Japan is presently leaning toward international responsibilities that do not involve a larger military responsibility, but of course this is no guarantee that it will continue to do so in the future. Japan has begun to increase its involvement in nonmilitary international activities. For several years prior to the Gulf War, Japan's foreign assistance had been increasing, and it is currently the largest aid donor in the world. Japan has demonstrated its support for international S&T programs. Japanese policy makers have also begun to turn the nation's attention more fully to addressing problems in the global environment. Should Japanese policy makers decide to become truly committed to this global responsibility and assume a serious leadership position, given America's not particularly impressive environmental record and the present uneasy state of the bilateral relationship, problems with the United States could easily worsen.

Some signs of Japan's understanding that it is a global superpower can also be seen in its recent trade discussions with the United States. At the same time, these discussions have demonstrated to the international community Japan's willingness to challenge the hegemonic leadership of the United States. The removal of the Liberal Democratic Party from power after holding it for decades soon made evident that change was imminent in the bilateral relationship. Japan was ready to establish a different relationship with the United States. During the short tenure of prime minister Hosokawa (August 1993 to April 1994), the Japan-U.S. Framework Talks collapsed only a few months after the establishment of a new bilateral policy structure. The fact that Japan did not adopt a concessionary posture in its discussions with the United States this time was unprecedented. Japanese policy makers, moreover, gave the strong signal that they were no longer going to yield automatically to American pressure in trade matters.

Although the departure of prime minister Hosokawa after only a few months in office and the subsequent protest withdraw of the socialists from the short-lived Hata government, which left it technically powerless, may have suggested some uncertainty in Japanese policy making, these events had no substantive impact on the bilateral relationship. While still experiencing pressure from the United States to 'adjust,' that is, to conform to American expectations, nonetheless a more autonomous Japan than had existed in the past was apparent. Led by prime minister Tomiichi Murayama, Japan again faced off with the United States in a new trade showdown in the spring of 1995, further demonstrating its resolve not to cave in to American pressure.

At this time American policy makers established a late June 1995 deadline for the implementation of severe sanctions on Japanese luxury automobiles because Japan refused to agree to set numerical targets on the importation of U.S. automobiles and parts. Despite America's key part in developing the rules of the newly formed World Trade Organization (WTO), the multilateral institution designed to judge and act on trade disputes and to replace the General Agreement on Tariffs and Trade (GATT), the United States moved decisively away from a multilateral settlement mechanism. Although the United States did register a complaint with the WTO, it also decided to employ unilateral action. Similar to the trade dispute in 1994 involving then-prime minister Hosokawa, Japanese policy makers made it perfectly clear that they were not about to concede to America's demands or be intimidated by its use of a unilateral stick. For the Japanese, relative to GATT the WTO greatly improves the multilateral mechanisms for settling trade disputes and accordingly reduces the prospect that a nation will take unilateral action. However, the United States feels that despite the existence of this new multilateral organization, unilateral action is still necessary. With regard to the 1995 auto dispute, the director general of the WTO indicated that the United States is 'no longer confident' of its leadership position in the global economy and is not ready to depend on a multilateral institution to resolve trade disputes.[23]

Only months later, Japanese reluctance to succumb to mounting American pressure was evident in the airline and film disputes. Most significantly, this reluctance was also apparent in the ongoing semiconductor arrangement. During the spring and summer of 1996 the government of prime minister Ryutaro Hashimoto demonstrated its refusal to concede easily to American pressure intended to extend the 1986 and 1991 Japan-U.S. semiconductor agreement, an accord designed to guarantee the United States a predetermined share of the Japanese domestic market. Japan not only welcomed settlement by the WTO in the 1996 semiconductor bilateral dispute,[24] but an additional matter connected to America's international security interests also had to be considered at this time. Because of the importance of semiconductors to both civilian and defense requirements and the U.S. desire to maintain a guaranteed share of the Japanese semiconductor market, which would help to offset high developmental expenses, Japan became more aware of the financial 'costs' connected to its alleged free ride; further, Japan's refusal to accede to U.S. demands suggests a temporary disassociation from America's concern with military preparedness.[25]

Japan and global security issues

Japan's a new international position, however, needs to become more comprehensive and focused on identifiable objectives. These objectives have to reflect Japan's national interests and those of the global community. In doing this therefore Japan's international position must not give the highest priority to American demands and interests. Japan's recent assertiveness with the United States in bilateral trade matters must be further developed and extended to areas that involve international issues and the global community.

This work will indicate a number of important policy mandates that have molded Japan into an advanced industrial nation that has rejected the use of military force to settle disputes with other nations.[26] Japan, moreover, has historically placed a great amount of faith in the U.N., and it has always preferred in principle to have this organization maintain global security. But Japan has not really taken a leading position within the U.N. because of its past dependence on the United States for protection and, accordingly, its willingness to follow America in matters pertaining to global security.

The historical convergence of two important developments has created the opportunity for Japan to work diligently toward the proliferation of the principles found within its constitution and that have become part of the way of life for a majority of the Japanese people. Japan's emergence as a global superpower and the end of serious ideological confrontation that had constantly threatened world peace provide it with the chance to assume a much more involved international position. Now determined to acquire a permanent seat on the U.N. security council, Japan can work actively to promote world peace and especially to engage global disarmament.

Japan's constitutional legacy makes it the most likely nation to pursue this as a major objective for itself as well as other nations of the world. Such an objective would clearly put Japan in an international position of unprecedented importance. Calls for Japan to be a civilian power in the post-Cold War period,[27] fall far short of recommending that it actively and dynamically take on the critical responsibility of promoting global disarmament.

Just as important for Japan is its relationship with the United States. As a hegemon that has 'stuck by its guns,' it is not very likely that the United States will endorse at any time in the near future a collective and permanent U.N. peacekeeping force that has complete responsibility for settling conflicts throughout the world.[28] But at the same time it needs to be pointed out that some of the bilateral friction has resulted from America's attempt to get Japan to assume more burden sharing responsibility.

It is true that the United States did acquire Japanese support in military production and technology development during the Korean War. However, continued American pressure on Japan regarding burden sharing began with the Nixon Doctrine in 1969. At this time the United States accused Japan of not paying its fair share of the military burden and attempted to get the Japanese to assume more military security responsibilities.[29]

Burden sharing has also been subtly tied to Japan-U.S. trade problems for several years. Implicit to the bilateral trade disputes has been Japan's security debt to the United States -- such as in the continuing semiconductor dispute. This therefore helps to explain Japan's trade concessions in the past, and since the end of the Cold War, its unwillingness to comply with

American demands, including those relating to S&T, without first fully considering its national interests.

One important way for Japan to circumvent pressure from the United States is to apply some of its own. By enthusiastically taking on a new responsibility that fully promotes global security and peace within the U.N., Japan would be in a position to put pressure on the United States, as well as other nations. This would not be retaliatory pressure, applied by one superpower on another in an unending power struggle. On the contrary, it would be a form of pressure which should ultimately improve bilateral relations, since in time the results should permit the United States (or at least provide it with the opportunity) to strengthen its industrial and S&T base with fewer claims of unfairness being made against Japan.[30]

American policy makers have long argued that many of Japan's acquired advantages in commercial markets have been the result of its relatively low military budgets over the years. To the extent that this is true, Japan can now make a very similar argument. But it must do more than this; to produce results, Japan must apply pressure on the United States.

Notes to chapter 1

1. See, for example, John Campbell, 'Japan and the United States: Games That Work,' in *Japan's Foreign Policy: After the Cold War, Coping with Change*, ed. Gerald Curtis (Armonk, NY: M.E. Sharpe, 1993), pp. 43-61.
2. Carol Carson, Bruce Grimm, and Carol Moylan, 'A Satellite Account for Research and Development,' *Survey of Current Business*, vol. 74, no. 11, November 1994, pp. 37-71.
3. Quoted in *Economic Report of the President*, (Washington, D.C.: U.S. Printing Office [USGPO], February 1995), p. 5.
4. Douglas Mendel, *The Japanese People and Foreign Policy: A Study of Public Opinion in Post-Treaty Japan* (Berkeley: University of California Press, 1961), pp. 94-102.
5. Anthony DiFilippo, *Military Spending and Industrial Decline: A Study of the American Machine Tool Industry* (Westport, CT: Greenwood, 1986).
6. See Anthony DiFilippo, *From Industry to Arms: The Political Economy of High Technology* (Westport, CT: Greenwood, 1990).
7. Ruth Sivard, *World Military and Social Expenditures 1993* (Washington, D.C.: World Priorities, 1993).
8. Eisuke Sakakibara, *Beyond Capitalism: The Japanese Model of Market Economics* (Lanham, MD: University Press of America, 1993).
9. On the topic of revisionist thinking in both Japan and the United States see Federal Research Division, Library of Congress, *Japan: A Country Profile*, eds. Ronald Dolan and Robert Worden (Washington, D.C.: USGPO, 1992), pp. 386-387.
10. The Clinton administration's conversion agenda is far from adequate. For example, see Anthony DiFilippo, 'The National Machine tool Partnership: Industry Savior or Band-Aid?' *The New Economy: The International Periodical on Economic Conversion and Disarmament*, Summer 1993, p. 7.
11. Center for Defense Information, Defense Monitor, *Policing World Trouble Spots: United States or United Nations*? vol. XXIII, no. 3, 1994, pp. 1-8.
12. William Tabb, *The Postwar Japanese System: Cultural Economy and Economic Transformation* (New York: Oxford University Press, 1995), pp. 22-28, 50-51.
13. See Ministry of International Trade and Industry, *Statistics on Japanese Industries 1993* (Tokyo: International Trade and Industry Statistics Association, March 30, 1993), p. 148; and Organization for

Economic Co-operation and Development (OECD), *OECD Economic Surveys 1992-1993, Japan* (Paris: OECD, 1993), pp. 64, 141.

14. See Research and Policy Committee for Economic Development, and a Statement by *Keizai Doyukai* (Japan Committee for Economic Development), *Japan in the Free World Economy* (New York: Committee for Economic Development, 1963).

15. Ibid.

16. See 'Publisher's Corner,' *Current Politics and Economics of Japan*, vol. 1, no. 1 (1991), p. 1.

17. Research and Policy Committee for Economic Development, *Japan in the Free World Economy*, pp. 40-41.

18. See, for example, Taizo Yakushiji, 'Technology and the Setting for Japan's Agenda,' in *Japan's International Agenda*, ed. Yoichi Funabashi (New York: New York University Press, 1994), pp. 57-80.

19. See *Japan Economic Almanac 1994* (Tokyo: *Nihon Keizai Shimbun*, Inc. 1994), p. 16.

20. For a current example of inter-government cooperation on involving industrial technology see 'Japan, U.S. Plan Joint Work on New Products,' *The Daily Yomiuri*, May 10, 1994, p. 8.

21. Science and Technology Agency, *White Paper on Science and Technology 1991* (Tokyo: The Japan Information Center of Science and Technology, 1992), pp. 226, 229-230.

22. See 'Japan, U.S. Firms Enter Microprocessor Pacts,' *The Nikkei Weekly*, May 2, 1994, pp. 1, 19.

23. See Subcommittee on Unfair Trade Policies, *1995 Report on The WTO Consistency of TradePolicies by Major Trading Partners* (Tokyo: World Trade Organization Committee of the Industrial Structure Council [an advisory body to the Ministry of International Trade and Industry], 1995); and 'U.S. and Japan Told to Solve Trade Dispute,' *The New York Times*, June 14, 1995, pp. D1, D5.

24. See, for example, Ministry of International Trade and Industry (MITI), *Our Position on the Japan-U.S. Semiconductor Arrangement* (Tokyo: MITI, December 1995), accessed on the World Wide Web.

25. Anthony DiFilippo, *'Handōtai Kyōtei: Encyō no Hitsuyou Nashi'* ('The Semiconductor Agreement: No Need for Extension'), *Nihon Keizai Shimbun*, June 22, 1996, p. 28. With funding by the Defense Advanced Research Projects Agency to the U.S. semiconductor consortium, SEMATECH, scheduled to end in 1997, developmental costs have become more important to American chip producers, who now can breathe a sigh of

relief that a new, albeit much weaker private-sector, bilateral agreement has been reached.

26. For a recent discussion of Japan's peace constitution see Charles Overby, 'A Quest for Peace with Article 9,' *Japan Quarterly*, April-June 1994, pp. 146-156.

27. Yoichi Funabashi, 'Introduction: Japan's International Agenda for the 1990s,' in *Japan's International Agenda*, pp. 1-27.

28. See 'New U.S. Peacekeeping Policy De-emphasizes Role of the U.N.,' *The New York Times*, May 6, 1994, pp. A1, A7.

29. Aurelia George, 'Japan and the United States: Dependent Ally or Equal Partner?' in J. A. A. Stockwin et al., *Dynamic and Immobilist Politics in Japan* (Honolulu: University of Hawaii Press, 1989), pp. 247-249.

30. See Anthony DiFilippo, 'Japan-U.S. Ties as a New Global Security Model,' *The Japan Times Weekly International Edition*, May 29-June 4, 1995, p. 8.

2 U.S. science and technology policy: development and objectives

This chapter is about the institutional forces that have shaped the science and technology (S&T) policy of the United States mainly since World War II.[1] Although other policy interests have certainly existed, America's national S&T policy from this time has been dominated by persistent efforts to make the nation the world's singular military superpower. This chapter also introduces the unexpected problem of how Japan, with help from America, moved with determination to use technology to rebuild its war-devastated economy, and further suggests that this was very important in putting it on an economic collision course with the United States more than a decade ago.

Background and context

The pre-World War II research efforts of the federal government concentrated mainly on nonmilitary work. For example, in 1937 and 1938 the War Department and the Department of Navy controlled about 20 and 22 percent respectively of regular federal research expenditures. Of the roughly 250 agencies having research potential, the largest actual performer at the time was the Department of Agriculture, spending about 31 percent of the total federal amount in both 1937 and 1938.[2] Also of interest is that government research spending was considerably less than that expended by industry. This was true from at least the 1920s until World War II. In 1940, for example, government (including state) spending on research was less than 30 percent of what industry spent, not much different from 1923.[3]

All of this changed after World War II when government's national research efforts shifted largely to maintaining a technological lead over what was previously the Soviet Union. Even though this formerly perceived threat no longer exists, the continuing desire for military superiority is still evident today. This means that despite the existence of nonmilitary interests, including commercial initiatives for industry, America's national S&T policy still has the same overriding military objective.

To speak of military ascendancy as the dominant objective in America's S&T policy is not to say that along with changing defense priorities there have not been other interests which have created occasional shifts in the nation's research and development (R&D) efforts. Though never enough to alter substantially the overall direction of S&T policy, the realization by policy makers that military and space (the latter for many years during the Cold War was an integral part of the nation's defense system) could not completely satisfy the diverse technological needs of the nation at different times led the appearance of civilian efforts that have had relatively low commitment levels by government. These nonmilitary efforts have always been designed to fill a perceived void that, even though urgent, the market has not sufficiently heeded for one reason or another; however, especially with civilian-industrial efforts, they have often been short-lived undertakings. Thus long-term planning has not been part of the government's characteristically *ad hoc* civilian-industrial S&T policy.

In the early 1950s U.S. policy makers successfully reconciled military-Keynesianism with the central precept of the classical school of economics, a noninterventionist government. We suspect that this occurred because the strong strain of individualism found in America could be more easily fitted to an activist government donning combat fatigues than civilian clothes, particularly since the effects of military spending served simultaneously as a perceived protector and as a short-term mechanism to help steer the economy. Moreover, the gratification gained from spin-offs from the military to the civilian sector, which during the 1950s and much of the 1960s were mainly perceived as adequate, provided both political and economic justification for resisting (but not eliminating) civilian-technology initiatives.

Not having a serious economic challenge for at least two decades after

World War II meant that policy makers never had to ruminate the virtues of a coordinated and extensive civilian S&T policy. Accordingly, U.S. policy makers continually apportioned large pools of capital to a military-space S&T complex, thus seriously minimizing possibilities for alternative uses. Few other policy decisions relating to directing the bulk of America's national S&T resources could have been worse for the industrial economy. Because substantial harm was imminent, this policy decision left the nation very unprepared to deal effectively with the rival industrial challenges that would eventually emerge from Japan, most especially. When this foreign economic challenge did appear, moreover, meeting it head on became increasingly difficult.

Significantly, after years the irony of U.S. policy making appeared: Japan's pacifist constitution, written with the help of American occupation officials, was a major contributing factor to strained bilateral relations. Japan's pacifist constitution put it on a S&T course quite different from that followed by the United States. America and Japan have not only been on different S&T 'trajectories' for decades,[4] but with Japan's ascension as an industrial and S&T superpower, the effects of these different paths have seriously collided from time to time on trade, technology, and military issues.

Hobson's choice -- or almost: military ascendancy in the early post-war years

In his 1945 report to the Truman White House, Vannevar Bush, then director of the Office of Scientific Research and Development, the agency which during World War II supported military research, made two fundamental points. As one of the leading proponents calling for the establishment of a National Research Foundation (the National Science Foundation), Bush made it abundantly clear in this report that the United States had 'no national policy for science' and that '[T]here is no body within Government charged with formulating or executing a national science policy.' Bush also noted that military research 'must be carried on in secret [and] much of it has no commercial value.'[5]

Bush called for additional and better defense research after the war, that is, during years of peace. However, it is unlikely that he anticipated that the military would become the main agency directing the nation's S&T policy. While Bush unequivocally recognized the noncivilian value of defense research, policy makers soon learned that spin-off rationalizations could provide increased public support for an expanding military-space complex.

From the Korean War through roughly the first half of the 1960s federal R&D spending was overwhelmingly for military and space purposes. Between 1953 and 1966 an average of 86 percent of federal R&D spending came from the Department of Defense (DOD) and the nation's space agency.[6] Most industry-supported R&D from the federal government (about 90 percent) was tied to these agencies. Thus during these years industries or organizations willing to do contract work with the federal government mainly got involved in military and space-related activities, for there was little opportunity to chose anything else. Federally supported civilian R&D spending paled in comparison to military-space expenditures; for example, in real (1982) terms military-space R&D totaled $24,001.4 million in 1960, compared to only $3,927.5 million in civilian expenditures.[7]

Nonetheless, real civilian R&D spending did grow from the early 1950s through the mid 1960s. (See Figure 2.1) While government R&D spending for military and supporting purposes came mainly from the DOD, the National Aeronautics and Space Administration (NASA -- before 1958 space and related S&T was directed by the National Advisory Committee for Aeronautics), and the Atomic Energy Commission (AEC), the budgets of other federal agencies also grew during this period.

It must be emphasized, however, that these civilian programs started from what were very modest beginnings. For example, the budget for 1950 supported the establishment of the National Science Foundation (NSF) and sought $2.5 million in funding.[8] Although its budget was relatively small, the NSF became the nation's premier agency for supporting basic civilian research. The Department of Health, Education and Welfare had an R&D budget of only $63 *million* by 1954, compared to $2.5 *billion* for just the DOD.[9] In 1961 government R&D funds obligated to transportation amounted to $55 million, and those for national resources and the environment came to $73 million, together, less that 2 percent of what was earmarked for national defense.[10]

Although the first half of the 1960s produced somewhat of an expansion in policy makers' civilian R&D interests, these concerns were largely confined to the Department of Health, Education and Welfare, the Department of Agriculture, and the NSF. Still, by 1966 the combined expenditures of these three agencies amounted to less than one fourth of NASA's spending and failed to exceed the AEC budget.[11]

Early on there was the push to sensitize the public to the civilian benefits

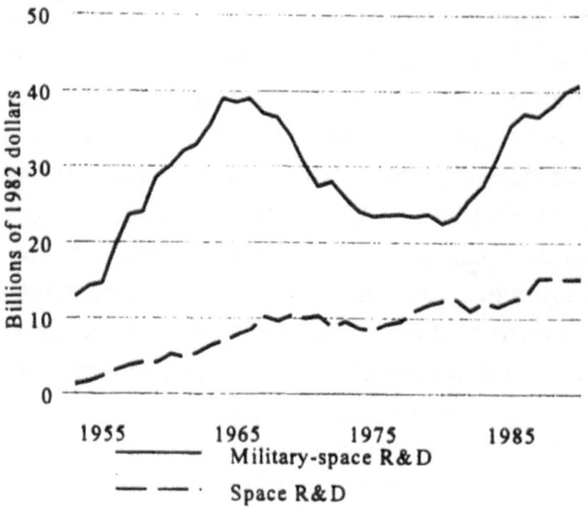

Figure 2.1 Real military-space and civilian R&D, 1953-1990

Sources: National Science Foundation (NSF) data; and price deflators from Economic Report of the President (1991).

flowing from military R&D. While the government acknowledged the preponderance of military and related R&D spending in the nation's S&T effort, there was the accompanying insistence that this work also had considerable civilian benefits.[12] One of the first government attempts to link military and civilian R&D was work supported by the AEC.

Established in 1946, the AEC was given the responsibility to develop nuclear technology for both military and civilian purposes. At the time the AEC was one of the few government agencies doing basic research, since the vast majority of government-supported R&D spending went to applied research. During the 1950s and into the 1960s AEC's military programs focused on developing technologies for nuclear weapons, and also for manufacturing them. Its civilian efforts were largely directed at building nuclear reactors, for the purpose of moving the nation toward widespread nuclear energy consumption. However, these reactors were to be used for both civilian and military applications.[13] Work on breeder reactors in the 1960s, which supposedly would 'produce more nuclear fuel than they consume' was therefore central to AEC's activities.[14] In 1965 the reactor program consumed almost two fifths of AEC's R&D spending, more than any of its other programs.[15]

AEC supported work on a number of nuclear reactors, including the one that powered the nation's first submarine in 1955. By the mid 1950s, as a result of the Atomic Energy Act (1954) and the implementation of its Power Reactor Demonstration Program (1955), AEC began to coordinate its work even more with the private sector in an attempt to give rise to a successful commercial nuclear power industry. Nearly half of the federal government's spending on nuclear power for commercial purposes through the 1950s came from military spending.[16]

The overlapping objectives of government to develop nuclear technology that would satisfy modern weapons requirements while also providing an increasing source of energy for the nation proved to be far too ambitious for the civilian sector. The failure to anticipate problems such as growing construction costs, lower demand for electricity than expected, the ever looming issue of public safety, and insufficient private investment in a risk technology, which resulted in the reluctance of industry to adequately engage in nuclear power, were enough to reduce significantly the prospect of success.[17] Add to this the government's putative military-civilian effort relating to nuclear power, which was part of a national S&T policy that gave the highest priority to defense and related R&D, and one has written the perfect formula for failure.

Apart from the big R&D expenditures of the DOD, AEC's R&D spending was much greater than that of any other government agency, including the nation's aeronautics program until 1963. Between 1954 and 1962 the R&D spending of the AEC was on average about 5 times greater than NASA's R&D budget. By 1963 the manned space flight program had become much more important to government policy makers. Indeed, R&D spending on the manned space flight program swelled from about $1.1 billion in 1963 to approximately $3.5 billion in 1966. By 1966 NASA spending on R&D was four times greater than that of the AEC.[18]

In the late 1950s government created the Advanced Research Projects Agency (ARPA), later, in 1972, to become the Defense Advanced Research Projects Agency (DARPA). Reflecting the very intense political tensions of the early Cold War period, U.S. officials perceived the need to bring advanced S&T into the mainstream of policy making. ARPA fulfilled this need.

ARPA formally began in February 1958. This was just four months after the Soviets successfully launched Sputnik I in early October of 1957. Sputnik II, accompanied by the Laika the dog, was launched in early November of 1957; preceding Sputnik I by six weeks, the Soviets had successfully tested an intercontinental ballistic missile in late August of

1957. While the appearance of ARPA remains the most visible manifestation of the Soviet successes with the Sputniks, it was not the only policy reaction. Scientists had managed to attract the attention of president Eisenhower, who reputedly, as one remarked, 'liked to think of himself as one of us.' Prior to ARPA's creation, in short, there was a dialogue between government officials and prominent scientists for the purpose of improving the connection between science and military preparedness. Resulting from this were two important institutional developments. During an address to the nation just four days after the launching of Sputnik II entitled 'Science in National Security,' Eisenhower announced the creation of the position of Special Assistant to the President for Science and Technology (the President's Science Adviser), and the President's Science Advisory Committee.[19]

ARPA's initial work mainly centered on space related research. Its fiscal year 1959 budget of $520 million provided a strong indication that government officials were prepared to fund ARPA adequately, permitting it to be financially prepared to develop advanced technologies. The creation of NASA in 1958 brought with it a major problem for policy makers: there was great difficulty in discerning the difference between military-space and civilian-space projects. In any case, by 1959 NASA had assumed most of the responsibility for America's space work, though essentially everything it was doing in the 1960s had appeared on ARPA's books in the second half of 1958.[20]

Motivated by the military's need for more resistant substances, an important early initiative of ARPA was financing the Interdisciplinary Laboratory Program which focused specifically on materials work at university research laboratories beginning in 1960. Though later viewing the results coming from the materials research work as insufficient, since ARPA's control over universities was limited by the block-grant distribution process, before 1961 was over it was funding eleven laboratories. Both AEC and NASA also implemented analogous initiatives at universities in the early 1960s.[21]

Although military and related S&T efforts reached into the halls of academia, it was U.S. industry that for more than a decade was overwhelmed by the technological largess associated with the political fervor of the Cold War. Indeed, excluding space spending, military and energy-related R&D expenditures were actually greater than those of industry until the second half of the 1960s, as we can see from Figure 2.2. Moreover, it is probably true that military and space R&D funds for

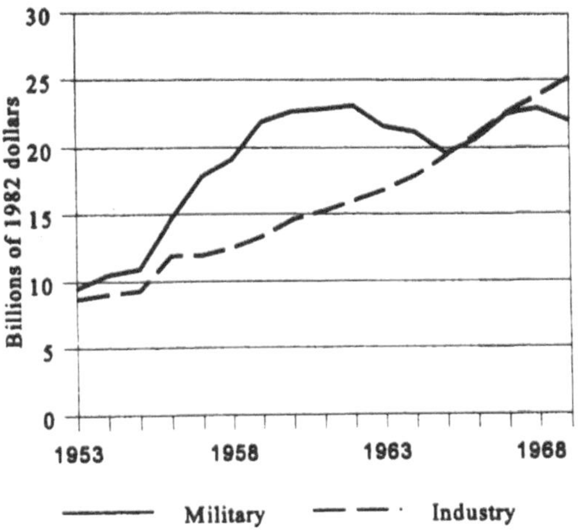

Figure 2.2 Real military and industry R&D, spending by source, 1953-1969

Source: National Science Foundation, National Patterns of R&D Resources: 1990.

industrial purposes during the early post-war years were close to the total expended for all civilian industrial development. Using the earliest available data on military and space R&D funds going to industry, we find that in 1963 spending by the DOD and NASA came to $6,500 million, and civilian-industrial R&D expenditures by the federal government amounted to $770 million. After adding the federal government's $770 million in 1963 for civilian-industrial purposes to company-financed R&D, which amounted $5,360 million, the nation's total R&D spending for civilian industrial development was still nearly $400 million below that devoted to military-space industrial work.[22] Given the dominating concern with military readiness in the United States during the early post-war years, it is reasonable to conclude that beginning in the mid 1950s and continuing until the second half of the 1960s federal R&D spending supporting work in the industrial sector, most of which was for defense and space, was very close to or *exceeded* company-financed industrial R&D.

Some civilian industrial R&D initiatives, however, did manage to receive public support, but only because of market neglect. One such enterprise was the supersonic transport project which began in the early 1960s. Because of the concern that the U.S. aircraft industry would not be able to keep up with the Soviet Union, France, and England in building a civilian supersonic vehicle, American policy makers turned their attention to this project in 1960. About a decade later, after having been federally funded from 1962, policy makers canceled the civilian supersonic transport. The project ended as a complete failure and cost the government nearly $1 billion dollars.[23]

In early 1964 six firms presented airframe and engine design proposals for the supersonic transport. Boeing and General Electric eventually won the competition, the former for the airframe and the latter for the engine. The original standards set by the Kennedy administration in 1961 stipulated that contractors would pay 25 percent of the cost and the contribution of government would not surpass $750 million. A further stipulation was that if contractors did not commit company funds or if economic problems were evident the project would have to be reevaluated. As it turned out, despite violations of these stipulations, the project continued until 1971.

Formed in 1958 and having both military and civilian functions, the Federal Aviation Agency (which in 1967 became the Department of Transportation) was responsible for the management of the project. However, the DOD did have significant input into the project during its existence.[24] The Deputy Administrator for Development in the Federal Aviation Agency testified during a hearing before a senate subcommittee in 1963 that: 'Without the wide base of technology created by the military programs, we couldn't even consider the SST [supersonic transport].'[25] In fact, the DOD and NASA worked with the Federal Aviation Agency on a two-year, $31 million precursory program to the supersonic transport project.[26] From the beginning contractors never liked the idea that they would have to bear 25 percent of the costs. When the contracts were let in 1967, Boeing and General Electric had to assume only 10 percent of the costs, and this did not include cost overruns.[27]

The program's failure can be traced to the contradictions that emerged between government and private sector contractors. In addition to the big technical problems of the supersonic transport, there was the shadowing political resistance that government should not be funding civilian projects. Suggesting this was that a funding maximum, although exceeded, was placed on the project from its inception.

This financial limitation, however, was only the surface of the problem. Of the six companies that entered the competition for the supersonic transport in 1964, five -- Lockheed, Boeing, General Electric, North American-Rockwell, and Curtiss-Wright -- were big military-serving firms. The sixth -- Pratt and Whitney -- had been awarded a sizable R&D contract from the DOD in fiscal year 1964 amounting to $612,000.[28] Thus even though the supersonic transport was a civilian project, key companies that had already experienced and benefited from DOD contract work entered the bidding process for the supersonic transport.

The successful contractors, Boeing and General Electric, had been among the biggest recipients of DOD contracts for some time. For example, between 1950 and 1957, Boeing was the second largest DOD contractor and General Electric was the fourth largest, with total awards exceeding $7 billion and $5 billion dollars respectively.[29] Because of their past experience with the DOD, Boeing and General Electric saw the government's proposed private cost sharing in the project as far too high. Using and risking company capital was not something these firms had become accustomed to when doing contract work with the DOD.

Boeing's experience with the DOD differed from that of General Electric's from the late 1950s to the end of the supersonic transport project. Unlike General Electric, Boeing's contract awards from the DOD had decreased considerably from the late 1950s to 1970. In 1958 Boeing was the leading military contractor, receiving awards that exceeded $2 billion; by 1970, the company had slipped to twelfth place, with DOD contracts amounting to only $474.7 million.[30] After correcting for inflation (in 1982 dollars), this meant that Boeing's military work had dropped drastically from $8.8 billion in 1958 to $1.3 billion in 1970.

When a House amendment passed that permitted the repayment fees to Boeing and General Electric for the use of private capital to go instead to the completion of a prototype for the supersonic transport during the project's final days, Boeing presented the government with an ultimatum: either it would receive an additional $500 million to $1 billion contract or it would discontinue prototype work on the supersonic transport.[31] Thus faced with declines in the company's contract business with the DOD, Boeing pushed policy makers to the wall, expecting a response similar to that experienced when doing defense work. The unwillingness of Congress to allocate the funds estimated by Boeing to salvage the supersonic transport translated into the termination of the project.

Constrained by the principle of limited government involvement in the economy and confronting the problem of Boeing's inability to keep costs down and to provide accurate cost estimates, the supersonic transport could not survive. The very serious issue of how to justify skyrocketing costs on a civilian project is fundamental to explaining the failure of the supersonic transport. While it is true that the supersonic transport was also becoming progressively less attractive for other reasons such as technical, environmental, and noise problems,[32] rapidly rising costs for a civilian-industrial project occasioned an inherent dissent among many policy makers. Given America's preeminent technological position in most areas, the failure of the supersonic transport therefore could be accepted by policy makers and even the business community.

More broadly, however, the preoccupation with developing the military-space R&D complex meant that U.S. policy makers could conveniently avoid seriously considering the implementation of a sound S&T policy for the civilian economy as a whole. Mainly for this reason, the United States missed opportunities to forge a government-industry relationship that would have advanced promising commercial technologies, many of which were mastered by the Japanese. One such technology was video tape recorders. Still another that came to be dominated by Japanese producers was liquid crystal technology used for screen displays.

U.S. technology diffusion -- target Japan

Damaged badly by the war, and both anxious and determined to rebuild its industrial base, Japan was very willing to import technology during the 1950s. Moreover, U.S. industry was equally willing to export technology to Japan. Between 1950 and 1959, according to the Science and Technology Agency of Japan, almost two thirds of its imported technology came from the United States. In 1963 Japan was paying more in royalties to the United States for its technology exports (licenses, patents, and advice) than it was to any other major industrial nation.[33]

Although U.S. companies found it very profitable to export technologies to Japan during the 1950s and into the 1960s, there was little investment in Japanese manufacturing during this time.[34] The lack of interest on the part of U.S. firms to invest in Japan, especially in the 1950s, was largely because of the poor performance of the Japanese economy. It is true that there was thorough scrutiny of foreign investments by the Japanese government, but these were legislated to prevent significant outside control of the economy. Through the Law Concerning Foreign Investment, which was passed in 1950 during the U.S. occupation, the Japanese government was able to guard against foreign control of the economy by limiting foreign investment and monitoring technology imports. These imports, which mostly took the form of licensing agreements with the United States, contributed greatly to the technology development of Japan's domestic economy. Japan not only imported U.S. technology, but because of these imports, it invested heavily in capital equipment. Data published by the Ministry of International Trade and Industry shows that between 1950 and 1960, technology imports by manufacturing firms contributed directly to almost 28 percent of the capital investments of companies. What is more, Japanese companies put these imported technologies to good use, upgrading them when feasible.[35] Japanese investment in R&D grew steadily relative to the United States. For example, between 1961 and 1966 Japan's R&D as a percentage of that of the United States increased from 8.5 percent to 10.6 percent.[36]

America's occupation of Japan meant that military spending there was nonexistent for a number of years. The American occupation of Japan was also responsible for the dissolution of the industry conglomerates known as *zaibatsu*, which the United States saw as a contributing cause of the war.[37] The military constraints resulting from the U.S. occupation helped Japanese policy makers bring in to perspective their effort to build a strong civilian economy. This, however, did not mean that the Japanese would not take advantage of the manufacturing opportunities made available to them from the Korean War.

Subsequent to the American occupation of Japan which ended on April 28, 1952, the influence of the United States on the Japanese economy continued to be quite strong. Following the U.S. occupation, moreover, Japanese policy makers made a choice not to rebuild the nation's military capability, instituting only a small Self Defense Force in 1954. Yielding formal military power to the United States, Japan's policy makers began crafting an economic policy that would focus almost entirely on growth. In Japan this policy was known as 'the separation of politics from economics,' one that effectively excluded the military from national economic activity.[38]

Thus by at least the mid 1950s Japanese policy makers had understood the direction in which they wanted the economy to head: Japan had renounced involvement in domestic military development -- a central part of its own heritage -- and put itself instead on a path of development that relied on continuous technological renewal. To accomplish this, the Japanese took in as much U.S. industrial technology as their economy could ingest. All of this was acceptable to U.S. policy makers for two important reasons. First, Japan's desire and commitment to follow a nonmilitary path meant that the United States would secure Japan and have increased military control over that part of the world. Secondly, U.S. business could profitably export industrial technology to Japan.

The Japanese made use of the American technology they believed would strengthen industry. The Japanese, however, did not stop there. Rather, they exploited this technology, making adjustments and improvements when necessary. Because most of the technology the Japanese purchased from the United States was still in the early stages of development, these adjustments and improvements were imperative. Determined to use technology to develop and broaden the industrial economy, the Japanese learned how to use to their advantage imported technologies in a wide range of manufacturing areas such as textiles, chemicals, electrical and nonelectrical machinery, metals, ceramics.[39]

Having signed two security treaties with the United States, one in 1951 and another in 1960, Japan had become incorporated into *Pax Americana*. Having no security alternative other than remaining unarmed and neutral as advocated by the socialist party,[40] Japan cast its security lot to the United States, despite a sizable amount of domestic resistance. There was, however, an important advantage for Japan, whether intended or not: it could continue to acquire corporate America's technology which was necessary for industrial growth.

But the bilateral relationship between Japan and the United States was destined for trouble. While Japan wanted to exploit foreign technology to strengthen its industrial base, the U.S. government had identified increased military competency acquired through the use of S&T as its priority. Gloating over the understanding that American technology was paramount, too many U.S. businesses were willing to sit back and permit government to assume too much of the technological risk -- a risk that centered far too much on advancing the military-space complex.

U.S. technology: policy and reality clash

The United States entered the second half of the 1960s with a technological monkey on its back; what is worse, it did not know it. Both government and industry continued to bask in the temporal reality of U.S. technological preeminence. Although rising real military and space R&D spending was no longer sustainable in a period of mounting social and political unrest, the decline and subsequent leveling off of these expenditures was not sufficient to sever industry's dependence on the government. Government's S&T policy for industry was subsidy-oriented and still largely for military and space purposes. To the extent that 'animal spirits' drive the engines of advancement in a capitalist economy, they could not count on government's trough for the S&T sustenance needed to propel commercial initiatives.

The S&T problems of the United States were not wholly the result of these developments, however. As Japan continued to make use of U.S. and other foreign technology, it also began to develop home-grown technologies by the mid 1960s.[41] The Japanese had found an industrial niche quite different from that of other industrial nations: by importing U.S. technological expertise and eventually by developing their own technologies they could move manufactured goods quickly to international markets. This was even true with some advanced technology products, especially in the case of the U.S. market. This meant that American technologies often boomeranged on the U.S. market in the form of imports from Japan. Moreover, as the United States turned more of its national energy to the Vietnam War, Japan was able to cash in on this by establishing a stronger position in the Asian market.[42]

In the United States the political dissension of the 1960s put increased pressure on government to turn more of its national S&T efforts toward social and political exigencies. Thus real civilian R&D by the federal government began moving noticeably upward in 1961. (See Figure 2.1.) The expansion of the welfare state created enough momentum to enlarge the R&D components of a number of social programs. Even the AEC joined in by aspiring in its Plowshare Program 'to develop peaceful applications of nuclear explosives.'[43]

R&D spending in a number of social programs began increasing in the late 1960s and continued into the 1970s. The R&D expenditures of the Office of Economic Opportunity amounted to $57 million in 1969 and grew by another $15 million two years later.[44] R&D outlays for Housing and Urban Development, while totaling only $12 million in 1969, grew to $48 million by 1973.[45] The Environmental Protection Agency spent $145 million on R&D in 1973; five years later, these expenditures had more than doubled.[46] Other social-civilian R&D programs that expanded in the late 1960s and into the 1970s were Income Security, Veterans Benefits, Community and Regional Development, and Education, Training, Employment and Social Services.[47] The successful effort by some policy makers to lessen the gap between R&D funds going to the military-space complex and those allocated for civilian purposes to accommodate growing sociopolitical and socioeconomic concerns is the most salient characteristic of government's S&T efforts during the period from about 1967 to 1981. The warfare-welfare problem that had manifested itself in political standoff of sorts had simultaneously established the parameters of the American government's S&T efforts. This, however, did industry little good, particularly given the policy aversion toward civilian-industrial endeavors.

Although the United States government continued to seek military competence and superiority through the commitment of what were still sizable S&T expenditures for these objectives, American industry went on the defensive. By the late 1960s the merchandise trade surplus of the United States had been reduced to about $600 million, down considerably from the $6.8 billion in 1964. Thus even though the trade balance improved in 1970, the 'shock' of America's merchandise trade deficit in 1971 was not really as unexpected as some analysts have made it out to be.

In any case, by the early 1970s there was an emergent, mostly rhetorical, concern in the Nixon administration that government needed to pursue a more comprehensive S&T agenda. Not yet in vogue, nonetheless, part of the administration's concern was with the decline of U.S. 'competitiveness' which it addressed by calling attention to some of its identifying components. Nixon's 1973 budget message purportedly was going to 'turn science and technology to the service of man.' A principal way in which this supposedly would be accomplished was by establishing a framework for the implementation of an improved industry-government 'partnership' that would generate new technologies and open up new markets. Other expected results of this proposed initiative were that both productivity growth and the nation's trade position in world markets would improve, as would employment prospects. To further enliven the technological capability of the civilian sector, the Nixon administration's 1973 budget called for the enlistment of the AEC and NASA to work on environmental, energy, and transportation problems important to the nation,[48] though, as we shall see below, not with the endurance needed for success.

Talk of changing the direction of America's S&T efforts during the midst of the Cold War was much easier said than done. Although real civilian R&D spending held its own in the first half of the decade, funds to the military-space complex were nearly three times higher. Guided by the arms control notion of 'stabilized deterrence' which countenanced increases in the number of strategic weapons in the U.S. arsenal,[49] Nixon's 1973 budget message also emphasized the intention to strengthen military R&D to make certain that the country would 'not face the possibility of technological surprise or lack of deterrent power necessary to protect' the United States.[50] Even in the latter half of the 1970s, that is, during the years of the Carter administration, when real civilian R&D spending grew somewhat as 'liberals' controlled the direction of the nation's S&T programs, military-space R&D expenditures were still more than twice as high as civilian spending.

The nonexistence of a commercialization policy in the United States meant that the optimal efficiency of America's national S&T resources were largely checked by the military-space complex. Once pulled into the military-space R&D complex, government S&T could only escape through occasional spin-offs. Moreover, faced with mounting political pressures for government to assist with domestic problems, policy makers could have easily identified additional areas where federal S&T spending could have been increased or where new applications could have been implemented. But a string of budget deficits created fiscal circumspection, and in the case of civilian-industrial projects, fiscal restraint.

This was evident in the case of research on magnetically levitated trains (maglev). Nixon's 1973 budget message specifically spoke of giving increased support through AEC and NASA to 'fast transportation systems.'[51] However, by 1975 the government had ended all support for work on maglev (ultra high speed trains). In fact, government support for research on maglev was never sufficient to move this technology very far along in the United States. Having gained a good deal of public attention in the United States during the late 1960s, maglev R&D received inadequate support not only from government but also from industry leaders, the latter not wanting to sink their own funds into this risky new technology.[52]

This reluctance of business leaders to use company funds on new initiatives and ultimately to commercialize them is also evident in screen display technology. In the beginning of the 1960s Westinghouse was seeking new ways to develop screen technology for improved vision. The most novel of these was thin or flat panel TV displays which, by using semiconductor innovations, had the potential to move toward a more advanced application known as active-matrix display. By 1967 Westinghouse executives began to lose faith in this innovative and risky work and threatened to cut off funding. When Westinghouse stopped producing TVs in the early 1970s, the R&D funding for thin panel displays was placed in jeopardy once again; this was followed by still another funding catastrophe a few years later when the company shut down one of its semiconductor divisions. In 1979 Westinghouse finally abandoned support for active-matrix display technology. As a result major Japanese companies such as Hitachi and Sharp were able to move more aggressively into this area. Currently the Japanese dominate the flat panel display market and completely control the active-matrix technology found in many flat panel screens.[53]

This story, however, is more complicated than just the failure of Westinghouse as well as other U.S. firms such as Sperry Rand[54] to continue developing advanced display technologies. Westinghouse experienced a major decline in military contracts during the fiscal period beginning in the second half of 1967. This is approximately the same time when Westinghouse first suggested it would stop funding thin panel display technology. The big cutback in Westinghouse's DOD contracts included a dramatic reduction (56 percent) in military R&D work at the time. By 1969 Westinghouse's military contract work had rebounded and for the next five years it maintained a relatively fixed place on the DOD contractors' list (ranging between positions 13 and 16).[55]

To the extent that Westinghouse saw the military market as at least somewhat secure, but potentially volatile, and since it had abandoned TV production at the time, the explanation of the firm's reluctance to support fully a risky consumer technology makes more sense. Between 1972 and 1973 Westinghouse's military contracts grew by 30 percent, including a more than $20 million increase in R&D awards.[56] As Japanese TV manufacturers acquired control of the market in the early 1970s,[57] and as Westinghouse was literally forced out of it, the exotic flat panel display technology was considered too extravagant to receive committed company support.

The erosion of the American TV industry ramified to budding industries such as high definition television and video tape recorders. In 1951 RCA started to do work on the development of the video tape recorder. After abandoning R&D on video tape recorders in 1958, RCA resumed this work in 1971. Along with Ampex, RCA had been involved in pioneering work on video tape recorders in the United States. By 1974, however, RCA again discontinued R&D on video tape recorders. Because a number of Japanese firms also initiated similar work during the 1950s, and through the persistence of government and industry, they were able to develop and eventually commercialized the VCR.[58] Not totally averse to taking risks, RCA decided to chance investing in the (very costly) VideoDisc, rather than stick with research work on video tape recorders. Given the deterioration of the American TV industry and the deep involvement of several Japanese firms in R&D on video tape recorders during the 1970s, the VideoDisc was a more attractive, albeit unsuccessful, investment for RCA.[59]

Similarly, the weakening of American TV manufacturing discouraged companies from moving toward research on high definition television (HDTV), and without any government-supported R&D program at the time, there was little reason for firms to be attracted to nurturing this technology. Like the Japanese, since at least the early 1960s U.S. manufacturers wanted to enhance the quality of TV picture images. But unlike the Japanese, U.S. manufacturers felt that the cost to undertake this type of technical problem was too big in an unidentified market. By the time the prospect of a HDTV market emerged at the end of the 1970s, U.S. manufacturers, then struggling in their efforts to hold declining shares of the TV market, were poorly situated to make any really serious move against the Japanese.[60]

Even before policy makers formally discovered America's competitiveness problem in the early 1980s, taking risks on new technological developments was not high on the list of priorities for most big U.S. companies. At least partly because the U.S. government never had a S&T policy for civilian industry, circumspection too often became a built in feature of the business mindset when it came to continuing innovative technology development. A bigger problem for U.S. manufacturers was that this was not the business philosophy employed by the Japanese, which by the mid 1970s had moved decidedly into a number of important markets.

One good example of this is maglev. Although research was first performed in the United States during the 1960s, the maglev concept quickly diffused to Japan and Germany. When government support for maglev ended in the United States in the mid 1970s, industry investment was not to be found; its interest in maglev was only rekindled through public spending.[61] Westinghouse, which drew up one of the first plans to build a maglev system, never followed through with the design.[62] Venture capitalists too were not attracted to maglev. This made maglev technology a prime target for foreign development. This especially meant Japan, which in 1979 tested a linear model that produced a speed of over 300 miles per hour (without passengers). The Japanese have worked diligently for years to develop maglev, relying on support from government. The Japanese plan to incorporate high temperature superconductivity into their work on maglev, a technology which is more advanced than the low temperature superconductivity magnet designs being tested now.[63]

America under siege

By the early 1980s there was an emerging realization that America's S&T

policy was not in keeping with the diverse needs of the nation. The importance of modernizing America's S&T policy, became a repeated topic of discussion both in and out of government. It was at this time too that analysts and policy makers alike 'discovered' that American business was not taking the risks necessary for an advanced industrial economy to thrive in the global economy. The main intellectual innovation that set the tone of domestic policy making on this issue during the early 1980s was that R&D and other long-term investments had been depressed because entrepreneurship had been crushed by the excess weight of government; the solution was for government to enact compensatory measures (i.e., tax initiatives) that would revitalize the supply side of the economy.[64]

It was also during the early 1980s that policy makers and analysts began to suspect that differences existed between the United States and especially Japan and that America's 'competitiveness problem' could be somehow traced to them. One main difference was Japan's approach to S&T. At the center of Japan's S&T programs was a government system that helped power the industrial economy. This was not only different from the political position acceptable to most U.S. policy makers, it was widely seen as wrong, for it violated a central tenet of free market (Western) capitalism. Rather than view the Japanese S&T efforts as a novel approach to inspire markets, American policy makers and analysts instead turned in droves to the more inclusive unfairness theme. After all, America's S&T supremacy was unmatched for decades. Given the nature of the paternalistic relation, it was generally reasoned that Japan was more like an errant adolescent that would eventually comply to political scolding and continued economic disciplining than an emerging S&T superpower. By linking Japan's different approach to S&T to its perceived willingness to practice unfair trade, policy makers and others had constructed what became an easily identifiable whipping boy for America's industrial and frequently technology-based problems.

Even though Japan had supposedly become an 'unfair' trading partner, and thus wrongfully acquiring comparative advantages in a growing number of markets, some policy makers and analysts nevertheless felt that perhaps America could learn a little from the successful S&T approaches practiced there. U.S. paternalism then was not without a liberal strain. Mainly it was understood that individual company funds could not match Japan's coordinated S&T system consisting of government, industry, public research institutes, and universities.

Reagan administration officials initially resisted even modest calls to introduce civilian changes in the nation's S&T program. America had gotten where it was through the ingenuity of the individual, the freedom of markets, and the willingness of companies to invest when things looked right. Following this line of reasoning, the Reagan administration successfully won the implementation of the R&D tax credit in 1981, followed three years later by the National Cooperative Research Act of 1984.[65]

The Reagan administration's most profound effect on the nation's S&T efforts, however, was through its military build-up. The build-up meant that companies would receive more DOD contracts, which would increase S&T activity. When this increase did not revive U.S. competitiveness, complaints of unfair trade practiced by the Japanese grew louder. Fundamentally, then, the years 1981 through 1986 was a retrograde period for American S&T policy making.

This type of intervention was rationalized by an ideological position which intimated that because military protection is a major responsibility of government, it was an acceptable intrusion into the free market. The upshot was that this not only displaced the focus from rejuvenated military-Keynesian policies, but it also carried with it the implication of regaining the economic prosperity of the past, since the private sector would accrue benefits through spin-offs.

Significantly, Reagan administration policies altered the narrowing gap between military and related R&D spending and that for civilian purposes. In 1981 the gap had at no other time since the early 1950s been narrower; by 1986 a considerable enlargement of this gap had taken place. While military-space expenditures grew rapidly, real civilian R&D spending leveled off between 1981 and 1986. To accomplish this, civilian R&D spending was thoroughly scrutinized for 'excesses' by the Reagan administration. In real 1982 dollars, R&D spending for education, employment, and social services declined by 200 percent between 1980 and 1983, that for community and regional development fell by 70 percent, and R&D funding for income security dropped by 43 percent.[66] R&D spending for other civilian purposes either declined less dramatically or remained relatively constant during this time; this was evident in transportation, natural resources and the environment, energy, and housing credit, as well as in federal support for civilian-industrial R&D work.

Faced with opposition in Congress, the Reagan administration was unable to sustain the momentum in the growth of real military and space R&D spending while neglecting civilian expenditures. By mid decade modifications were made in America's S&T policy. Real R&D spending for military and space purposes declined modestly in 1987, while funds for civilian purposes increased slightly. The decline in real military and space R&D was only temporary, however. Leveling off for the remainder of the decade, real civilian R&D spending by the government gave way to the technical interests of the military-space complex.

The rhetoric that government had no place in the movement of market forces for industry notwithstanding, by 1985 the Reagan administration began introducing new civilian-industrial S&T initiatives -- thus acknowledging the crisis in America's S&T policy. The institutional conduit with which it did this was the Commerce Department's National Bureau of Standards (later, at the end of the 1980s, to become the National Institute of Standards and Technology). Civilian R&D spending increased for innovative efforts relating to 'manufacturing process and quality control and biotechnology.' R&D money was also made available for the on-site building of the Cold Neutron Research Facility for work on materials science.[67] U.S. industry had demonstrated weakness in materials science and it was particularly the Japanese who had made impressive gains in this area, which, for example, supports the electronics industry.[68] The Bureau also received increased civilian funding specifically for work that would strengthen 'optic fiber systems to support industrial development and to ensure international competitiveness.'[69]

What is evident is that America's S&T system confronted a series of unprecedented conditions beginning in the mid 1980s; in its existing form, it was, simply put, unable to meet the foreign challenge. Moreover, the results from the billions spent on military and space R&D was not finding its way to the civilian sector with anywhere near the frequency needed to stave off competition from overseas producers. This institutionally restrained diffusion prompted the Reagan administration to propose in its 1988 budget the more expeditious transfer of R&D work performed in government laboratories to the private sector.[70]

During the second half of the 1980s policy makers also saw the need to increase support for basic civilian research that linked to industry. The NSF became an important part of this effort. Building on the idea of strengthening engineer training at universities through NSF's Engineering Research Centers Program, policy makers launched the Science and Technology Centers Program in 1988. Similar to the Engineering Research Centers Program, the NSF's Science and Technology Centers Program is a university-located and training-oriented initiative that includes the private sector, as well as state and local governments. This program, however, is for the training of both engineers and scientists and includes support for scientific research in a number of fields such as materials and the communication and information sciences. The NSF established eleven centers in 1989, with program funding amounting to $150 million.[71]

What had become evident is that the point had been reached where America's S&T policy, despite the ample padding of the military-space R&D complex, had shown itself to be inadequate in an environment where especially Japan, a new industrial superpower, had emerged as a genuine challenger to the United States. This forced U.S. policy makers to take steps (the modest funding of advanced S&T work for civilian-industrial purposes, along with managed trade), even though they were completely out of line with an administration that avowed to be fully committed to the free market.

However, this was not all that different from the S&T course the United States had followed in the past. Impending or actual crisis had pushed policy makers into action before. The Kennedy administration responded to the crisis in civilian air travel, as we have seen, with the supersonic transport. The Nixon, Ford, and Carter administrations supported civilian R&D during the period when America confronted an energy crisis. Indeed, the Department of Energy was established in 1977 during the Carter years for the purpose of advancing energy technologies.[72] Like their predecessors in the Nixon and Ford Administrations, policy makers in the Carter White House supported R&D work involved in producing synthetic fuels from coal, and nuclear technology (especially the liquid metal fast breeder reactor) with both military and commercial potential, though they did not endorse the Clinch River Breeder Reactor because they saw it as an untimely demonstration project that would not yield enough economic benefits.[73]

Carter administration policy makers, along with Congress, had a growing interest in solar technologies. Shaken by the energy crisis, Carter administration officials felt that it was government's responsibility to invest in technologies, such as solar power, since industry was unlikely to make the necessary long-term financial commitment. Special attention was given to photovoltaics, a technology that converts the sun's energy into electricity. (Reagan policy makers reduced by more than half spending for solar energy R&D because they preferred to fund projects that were more of a risk for the private sector and required a long-term commitment.)[74] The Carter administration also implemented a new cooperative government-industry research project for the automobile industry. Carter's 1981 budget proposed funding this cooperative venture from an energy windfall profits tax. Carter's 1982 budget showed continued support for this project; it proposed that an estimated $17 million be allotted for basic automotive research through the Department of Transportation, with an equal amount coming from industry.[75]

What the Reagan and Bush administrations did do differently from those that preceded them was that they were more publicly insistent that government should have very limited involvement in commercial S&T affairs. For example, the Bush administration maintained that there are 'very real limits on what the Federal government should do to promote technological success: the private sector must carry the responsibility for developing commercial products. The government role is to foster a stable economic environment that is conducive to investment.'[76] However, S&T initiatives in the Reagan-Bush White House noticeably departed from this position.

With mounting S&T-related challenges coming from Japan (which included significantly improving manufacturing efficiency), the Reagan and Bush administrations were forced to support commercial initiatives. One of the things the Reagan policy makers did consistently in the second half of the 1980s was to reduce overall R&D funding for the Department of Commerce while increasing obligations for certain civilian technologies (see above) within the department's National Bureau of Standards.[77] However, an important point must be emphasized: as was true during the Reagan years, support from the Bush administration for civilian S&T initiatives was both necessary and reluctant. This showed up, for example, in the debate over government support for HDTV. The Bush administration was reluctant to support HDTV, even through DARPA, because it appeared as if the government was 'picking a winner.' Helping to circumvent criticism and thus deal with this reluctance was the changing of the name to high definition systems.[78]

Where has all the security technology gone?

During the Reagan and Bush administrations there still existed the notion that the effects of military and related spending would be a catalyst to the nation's stubborn productivity and innovation problems.[79] Like in the past, military and related R&D would create the potential of technological fallout for the civilian sector. A further development was that NASA was given more of a 'civilian' look, particularly through its Commercial Development of Space Program begun during the Reagan years.[80] With at least 16 Centers, this program was supposedly established to foster a cooperative relationship between state governments, businesses, and universities interested in the commercial uses of space technology. In 1989 NASA's budget for commercial programs, however, made up less that 1 percent of total obligations for R&D.[81] Nonetheless, NASA remained involved in an effort to demonstrate the commercial value of its work.

By the mid 1980s it was becoming obvious to some U.S. policy makers and to other observers as well that despite America's sizable investment in military-space R&D, it was not doing the job of both securing the nation and stimulating industry enough to keep it acceptably competitive. The question policy makers confronted was unprecedented: why were not the significant increases in (real) military-space R&D spending and the many new, advanced defense technologies yielding satisfactory results to civilian industry?

The answer to this question has three different parts. In time many interested policy makers and analysts came to accept the conclusion that even beyond the military technologies that could not be commercialized, the combined impact of bureaucratic and technological strictures had too frequently moved much of the military-space milieu too far from commercial objectives. This explains the sustained effort to transfer government supported R&D to usable commercial outcomes -- introduced in the Federal Technology Transfer Act of 1986 and subsequently, in 1990, appearing as part of *official* U.S. technology policy.[82]

Second, although policy makers understood that Japan had become an important manufacturing and technological force during the 1980s, they largely viewed this as independent of trade-related matters. For whatever reasons, they did not pay enough attention to the significance of Japan's industrial and technological emergence, specifically in regard to its *relative* effect on short-circuiting U.S. commercialization efforts. As a result, because of where most of the federal government's R&D funds were going, it was mainly the DOD that had to step in and support some important technology areas such as semiconductors and high temperature superconductivity.[83]

Third, as the satisfied recipients of product and R&D contracts from the DOD, military-serving firms never had much of an incentive to expedite the commercialization of technologies, even when this was feasible. During the 1980s this was further complicated by two things: rapidly growing DOD contracts, and the tendency for military-serving firms to 'wall off' their civilian divisions from their defense-producing sectors.[84] This occurred at precisely the time when the Japan was moving to secure its position as a technological and industrial superpower.

Dual-use technologies

Increasing Japanese competency in manufacturing and the simultaneous development of technologies having both civilian and military applications had far-reaching effects on U.S. defense and S&T policies. Indeed, this led the Japanese to speak of technologies that 'spin on' to the defense sector. The very idea of technologies spinning on from the commercial to the defense sector fundamentally indicated that U.S. military preparedness could not rely entirely on home grown technologies. For some observers, by at least the mid 1980s the fear of growing foreign technology dependency became an important issue. At the same time, the shortcomings of America's S&T efforts in conjunction with the competitiveness problem caused increasing attention to be paid to dual-use technology.

The popularized notion of dual-use is that technologies are developed irrespective of application in the military or civilian sectors. Since a significant change in America's S&T policy was not considered an option during the 1980s, by the latter part of the decade dual-use technology had become an integral part of the federal government's S&T agenda. The formalization of dual-use policy suggested several important points. First, it substantiated that the development of civilian technologies had in too many areas pulled ahead of those in the defense sector. Second, because the DOD revealed that foreign technology dependence had become a concern of the military, it could be rightly concluded that at least some foreign technology had become more advanced than that developed in the United States. Third, the introduction of a dual-use policy tacitly suggested that spin-offs from the military sector were inadequate for America's defense needs, and were therefore not sufficiently available to keep U.S. industry competitive. Science adviser to president Bush, Allan Bromley, more than alluded to much of this when he remarked: 'There used to be substantial spin-offs from the military. Now, the military draws more and more from the consumer side.'[85]

Parrying suggestions that a civilian S&T agency be established by the federal government, U.S. policy makers toward the end of the 1980s gave the DOD the responsibility of identifying technologies that would foster national security. However, there is an important change to be noted here: national security took on a broader meaning by the end of the 1980s; it no longer conveyed the singular message of military security. In the past national security was synonymous with military preparedness. By the late 1980s, however, policy makers and most analysts had added a new dimension to the meaning of national security -- economic growth and international competitiveness, which was soon linked to S&T.[86]

Foreign technology belonging to Japan was becoming increasingly important to the United States. DOD estimates of Japanese industrial technologies dating as far back as 1984 identified 15 dual-use technology areas where the Japanese had either pulled ahead of or reached parity with the United States.[87] More recent assessments by the DOD suggest that America's military has grown more dependent on the technological advancements of the Japanese. The DOD's 1991 list of the 21 technologies which it viewed as crucial to national security identified only 6 that were mainly defense technologies, though between fiscal years 1987 and 1991 they made up 25 percent of the spending on critical technologies. The rest, said the DOD, are dual-use technologies. Even in a few of the mainly defense technology areas, the DOD viewed Japanese technological advancements as appreciable such as in the majority of sensitive radar technologies, and in pulsed power.[88] Moreover, according to one of Japan's leading politicians, many of America's advanced weapons systems would not be operational without his nation's semiconductors.[89]

Besides semiconductors, the DOD indicated the possession of a high level of sophistication by Japan in other areas such as biotechnology, specifically, bioelectronics and biosensor technologies, in composite materials, especially ceramics, and in superconductivity, particularly high temperature superconductivity. In machine intelligence and robotics, the DOD similarly acknowledged the technological proficiency of the Japanese. Information technologies have also been pushed forward by the Japanese. This is especially true in photonics, the technology that relies on light to pass and control information. Japanese technical accomplishments in photonics are very notable in fiber optics.[90]

While the 1989 fiscal year Defense Authorization Act mandated the DOD to develop an annual critical technologies plan, the next year this act created the executive office's National Critical Technologies Panel. Producing a biennial study to the year 2000, the Panel has a membership selected by White House's Director of the Office of Science and Technology Policy. In

1991 the Panel identified 22 technologies 'deemed critical to the satisfaction of national needs.' This list, along with a similar one constructed by the Department of Commerce in 1990, 'demonstrates the substantial degree of overlap that exists between those technologies essential for national security and those that contribute to economic competitiveness.'[91]

Thus dual-use became the single most important policy concept in America's S&T inventory. The problem, however, was not conceptualizing dual-use, rather the difficulty was (and remains) effectively implementing it. Parts of the academic community, accordingly, attempted to address this problem through the endorsement and propagation of the dual-use policy.[92]

Whereas in the past military-space technologies were said to be the catalyst of America's technological infrastructure, now the technologies of the civilian sector, which were being increasingly incorporated into military work, had somehow pulled ahead of those in the defense sector. Indeed, this 'pulling ahead phenomenon' is the most fundamental assumption of the dual-use approach. If it were not, there would be no reason for policy makers and analysts to abandon the passive (i.e., with respect to the civilian sector) spin-off mindset.

Just how the civilian sector pulled ahead of the military sector in the United States, however, has never been satisfactorily explained by supporters of dual-use policy. A major argument supporting this position is that during the 1980s, unlike in the previous decades, company expenditures for R&D were larger than those of the government.[93] But when looking specifically at industrial R&D supported by government and that financed by the private sector, companies have been outspending the federal government for years. Each year since 1968 R&D expenditures by companies have surpassed industrial R&D spending by the federal government. As time passed, the gap grew wider. Making this argument even more dubious is that during the early post-war years when federal support of industrial R&D was greater than company spending (see Figure 2.3), American industries unquestionably dominated global markets.

According to the argument supporting the dual-use position, innovations from the civilian-industrial sector should have pulled ahead of the military sector long before the mid to late 1980s, even after considering a time lag from the effects of technology to the industrial economy. Making the 'pulling ahead phenomenon' less credible is that this transition in the source of R&D expenditures, which began in 1980, was occurring at the same time that U.S. competitiveness was being continually weakened by aggressive

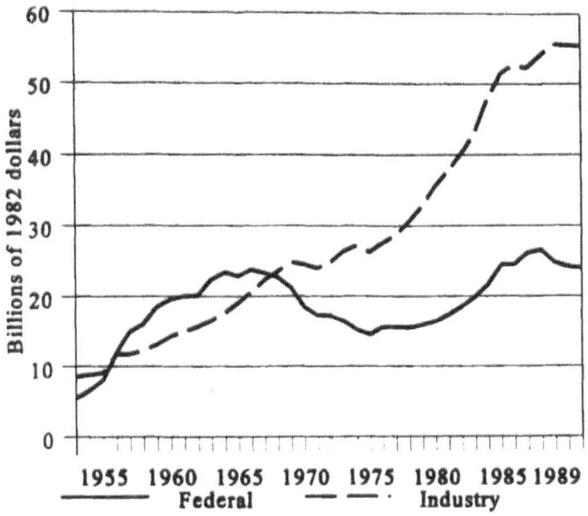

Figure 2.3 Real industrial R&D, federal and company, 1953-1991

Source: National Science Foundation, Selected Data on Research and Development in Industry: 1989 (Washington, D.C.: NSF, February 1991).

foreign industry. If this was really a new beginning for American industry, should not some of the commercially inspired technologies developed from this R&D transition have found their way to the market place during the 1980s and improved the nation's international competitiveness?

The civilian sector did indeed pull ahead of the military sector. However, it is misleading to suggest that it was the U.S. civilian sector alone that was responsible for this accomplishment, particularly during a period when there were major technological developments in other nations, most notably in Japan. Table 2.1 shows that from 1981 to 1988 the ratio of Japanese to U.S. company-financed R&D performed by businesses grew substantially. This data indicates that company-supported R&D has increased more quickly in Japan than in the United States. Moreover, this trend dates back to at least 1965 when the ratio was much lower (0.16).[94]

What does all of this tell us? Our position is that it is inaccurate to suggest that technical innovations in the U.S. civilian sector have pulled ahead of those in the military sector without considering Japan's contributions. As we will discuss in greater detail in Chapter 5, the recognition by proponents of dual-use that the civilian sector has moved ahead of the military sector needs to be interpreted in a much broader context than it is by U.S. observers. Japanese research and technologies have become sufficiently

Table 2.1 Company-financed research, Japanese to American, 1981-1988

Year	Percentage
1982	42%
1983	44%
1984	44%
1985	46%
1986	45%
1987	47%
1988	51%

Source: Organization for Economic Co-operation and Development (OECD), Basic Science and Technology Statistics (Paris: OECD, 1991).

advanced to warrant ongoing efforts by the United States to have access to this work. The issue of American access to Japanese technology became important to the United States during the early 1980s. By the second half of the 1980s, therefore, both dual-use and what soon became known as 'symmetrical access' had become very important policy issues for the United States. Both were becoming increasingly important at the same time that trade relations between the United States and Japan were in a state of obvious deterioration. Moreover, both required that government become much more active in the nation's S&T efforts than it was in the past. Most significant, is that both dual-use and symmetrical access are intended to strengthen America's national security by broadening the availability of military and civilian technology development.

The dual-use approach has permitted many policy makers and analysts to reconcile their resistance to government intervention in S&T enterprises because of the argument that this form of public support has both military and civilian applications and typically because of the so-called 'generic' or 'precompetitive' nature of the work. Precompetitive research is work that falls anywhere along the R&D spectrum -- from basic, to applied, to developmental -- and in the United States is said to be nonproprietary.[95] Because dual-use did not require significant changes in policy making, it became the chief instrument used in the effort to modernize the federal government's S&T initiatives.

This became evident during the years of the Bush administration. Dual-use had become a policy imperative; it was so important for America's national security interests that the Bush administration had to reconceptualize government's role in supporting civilian work. Because it incorporates generic and precompetitive research and involves military readiness, government-supported dual-use initiatives were rationalized as legitimate ways to assist private sector work. Specifically, the Bush administration took the position that government could fund applied civilian research by supporting the work of agencies in areas 'where some of the R&D has broad applications in the private sector ('dual-use' technologies), even though these applications would not necessarily be funded by the government.' Although against the targeting of specific industries, the Bush administration rationalized that 'the Government's role is to support generic or enabling technologies at the pre-competitive stage of R&D.'[96]

However, the industry-related S&T crisis that had emerged in the United States was too important to permit policy makers to lay back on their heels. The push to widen the meaning of national security became even too much for the market purists within the Bush administration. A telltale that by the early 1990s national security had taken on a broader meaning than in the past, and so incorporated a definitive economic dimension that directly related to competitiveness, was that even civilian-useful technologies such as maglev, photovoltaics, and electric and hybrid vehicles began receiving funding from the government.[97]

The Clinton administration stepped-up government's support for dual-use technology. Its advisory team brought to government the view that because of weakened American competitiveness, the movement toward public support of 'conversion technologies' had merit. This, concomitantly, increased the political attractiveness of dual-use in Washington. The end of the Cold War, of course, made it easier to promulgate this kind of 'economic reform' and to acknowledge freely the broadened meaning of national security.

Clinton himself relied on this policy approach even before his election. Resting on the realization of the Cold War's end and recognizing the incontrovertible importance of S&T to promote economic growth, the core of his proposed strategy was to call for a gradual shift back to where half of government's R&D budget would go to defense and the other half would be spent on nonmilitary efforts.[98] The reasoning was that because the political period legitimated, and indeed demanded, such a shift, a global high tech military presence could still be ensured even by proportionally reducing defense R&D. Moreover, civilian-related S&T supported by the federal government would benefit from the tradeoff -- and this essentially was an acknowledgment that industrial competitiveness had become a recognized part of the meaning of national security.

To accelerate the dual-use paradigm, the Clinton administration gave immediate attention to DARPA. Early in 1993 the Clinton administration increased DARPA's national security responsibilities and gave it a name change: it was renamed the Advanced Research Projects Agency (its name prior to 1972), and plans were made to enlarge ARPA's dual-use activities. This purportedly would 'increase the likelihood that defense research can lead to civilian opportunities.'[99] The ARPA name did not stick; the 1996 fiscal year Defense Authorization Act mandated that ARPA once again become DARPA.

In March 1993 the Clinton administration announced its program on Defense Reinvestment and Conversion. This five-year program combines the efforts of four government agencies -- DOD, DOE, the Department of Labor, and the Department of Commerce. Funding for this program increased from almost $2 billion in 1993 to approximately $4 billion in 1994, with an additional $800 million proposed for 1995.[100] The U.S. government's interest in defense conversion has currently become noticeably less enthusiastic than it was just a few years ago.

An important part of the Clinton administration's new conversion program is the Technology Reinvestment Project. Given legal authority by the Defense Conversion, Reinvestment, and Transition Assistance Act of Fiscal Year 1993, as well as other laws, the Technology Reinvestment Project (TRP) came into existence. Managed by the Defense Technology Conversion Council, which is led by DARPA, TRP involves four other government bodies, DOE's Defense Programs, Commerce's National Institute of Standards and Technology, the NSF, and NASA. TRP (a cost-sharing program with participants required to supply a minimum of half the total funding) received nearly $500 million through the Fiscal Year 1993 Defense Appropriations Act.[101] The proposal for TRP in 1995 was that funding would exceed $600 million.[102] (This subsequently changed; see below.) The main purpose of TRP is to develop dual use technologies. This means that all TRP projects must be demonstrably relevant to the Department of Defense.[103]

Can dual-use work?

Although dual-use is now part of the official S&T policy of the United States, it directly confronts big problems which impede its chances for success. Even for those not already troubled by the DOD leading national S&T development in the United States, these problems are cause for serious concern. Identified below are four reasons why it will be very difficult for the dual-use paradigm to work effectively, that is, not in itself being responsible for markedly improving America's commercial S&T base and therefore restoring the weakened competitiveness of industry.

First, proponents of dual-use assume that through it S&T development in the United States will become stronger *and* that the military's technological needs will be liberated from foreign dependence.[104] This is an unrealistic assumption since it does not address the relative difference in the growth of U.S. and Japanese industrial R&D expenditures; nor does it consider the nurturing civilian-industrial efforts undertaken by the government of Japan.

Second, U.S. companies that have performed contract work for the DOD typically have separate military and civilian divisions. Rarely, if at all, do they do business together. This means also that the S&T components of these divisions are separate from each other.

Breaking down these 'walls' will be no easy task. Companies have built them to minimize as best they can the problems associated with performing defense and related contract work. Their absence would create added economic and technological inefficiency and exorbitant costs for civilian-producing sectors. Civilian sectors of military-serving firms then have no sound reason to be more exposed to the problems of military divisions. Suggestions made to break down these walls by encouraging more cooperation between sectors and through reduced DOD administrative control omit a key issue: because the U.S. military is now more than ever a definitive technological system, there still remains the need for concealment. This remaining concealment requirement is a major stumbling block standing in the way of S&T 'fluidity' between the defense and civilian sectors.

The third problem related to the unlikely prospect of success for the dual-use paradigm is how to draw into the military and associated network civilian companies that are working on advanced technologies but prefer not to be a part of it. Some civilian companies are reluctant to become associated with this network for fear that their ability to compete in international markets will be compromised by defense-related contract work. Because dual-use work performed in the United States generally has defense relevance, and since dual-use is a policy which is inherently restricted by its intimate connection to the DOD, some companies will not want to become involved in contract work of this sort.

Fourth, dual-use spending by the federal government is only a very small part of the Defense Department's R&D budget. With just a little over $2 billion budgeted for military dual-use in 1995, these technologies made up no more than 5 percent of the DOD's budget for R&D.[105] This hardly seems like a sufficient amount to make American industry more competitive in areas involving advanced S&T.

We must also stress that the dual-use paradigm has a circular and, what is worse, an unconvincing logic connected to it. This makes it considerably more of a political rationalization than an effective policy instrument. Because the expectations are that 'potential future commercial applications' of applied civilian R&D will improve defense readiness, they fall into the dual-use category, and so can be justifiably supported by the federal government, as of course can defense technologies that have potential civilian applications. The political thread that pulls together defense and civilian efforts is national security -- broadly defined. Thus dual-use is the technological equivalent of 'killing two birds with one stone,' the expectation being that success is merely contingent on sectorial adaptability. In the United States, however, it has yet to be demonstrated that enough fluidity exists between the military and civilian sectors, particularly with the expeditiousness required to attend to the commercial and competitive demands of national security.

Some recent developments in America's S&T policy

As we have seen, during the 1980s it became clear that America's S&T policy was in a state of crisis. Consequently, a number of efforts, some of which rely on the dual-use strategy, have been undertaken to ameliorate the efficiency of America's S&T infrastructure. Others such as SEMATECH (semiconductors) and the Critical Technologies Institute (machine tools) took aim squarely at specific industries.[106] All have one thing in common: they are designed to forestall the further erosion of the nation's S&T infrastructure, and in the best case scenario, to strengthen its competitiveness. They can not, however, be expected to substitute for a coordinated and comprehensive S&T policy that focuses specifically on commercial growth -- a policy that has the attendant concern with job growth and decent employment.

In November of 1993 the Clinton administration started the National Science and Technology Council. This is a cabinet-level office with a mission to coordinate federal S&T. Begun also at this time was a private sector advisory committee which has the role of advising the council and the president.

The Commerce Department's National Institute of Standards and Technology (formerly the National Bureau of Standards) has been becoming increasingly popular in Washington during the past several years. Two of its programs, the Manufacturing Extension Partnerships (MEP), and the Advanced Technology Program (ATP), have been frequently identified as being especially promising initiatives for S&T development.

Conceived by Congress in 1988, ATP received $10 million in 1990. Because it moved government unequivocally into civilian R&D areas with the intention of nurturing and quickening commercialization, ATP had been billed by the Bush administration as an 'experimental' program involved in 'industry-led high risk R&D on pre-competitive, generic technologies.'[107] Part of the Bush administration's justification for government support of potentially relevant civilian technologies was the use of the terms generic technologies and pre-competitive research. As a former Commerce Department deputy secretary in the Bush administration pointed out, together, the use of these terms were 'an artifice to keep us from being destroyed.'[108] Funding for ATP has been noticeably increased by the Clinton administration. Funding for ATP in 1995 was about $450 million, over twice the amount allotted in 1994; ATP expenditures also increased in 1996.[109]

Also seen as 'experimental' by the Bush administration was the National Institute of Standards and Technology's Manufacturing Technology Centers. In 1992 the Bush administration allotted $10 million to this program, which consisted of six centers.[110] The institute now supports a number of centers through its MEP. Funding by the Clinton administration for MEP was in the neighborhood of $60 million in 1995, about double what it was in 1994.[111]

There are still other major S&T initiatives that have appeared in recent years. One of these is the cooperative research and development agreement (CRADA). The National Institute of Standards and Technology and some federal laboratories have been able to initiate CRADAs with industry and other nongovernment organizations since 1986, but those involving the

DOE were not permitted until 1989. The National Competitiveness and Technology Transfer Act of 1989 gave the DOE laboratories the authority to participate in CRADAs. Although DOE CRADAs got off to a slow start, by the spring of 1993 the number had reached 382, with funding amounting to $321 million. However, there have been complaints that DOE CRADAs are not reviewed and implemented expeditiously, and that the department's defense program CRADAs, which receive the majority of the funding, are especially slow at this. CRADAs with the DOE's weapons laboratories, which take at least eight months from the time of proposal submission to implementation, have created much industry frustration.[112]

The end of the Cold War and the need to increase industry's competitiveness have created the imperative for the transfer of technology from government to the private sector. For these reasons, federal funds for technology transfer have risen rather sharply in the past few years. Funding for technology transfer exceeded $500 million in 1994; spending for this program approached $2 billion in 1996.[113]

Then and now: has the primary objective changed?

The most important thing that stands out in this analysis of the S&T policies of the United States over the past five decades is that the federal government has never been fully committed to commercialization and industry competitiveness. During most of this time, commitment to commercial initiatives by the federal government has been tepid at best. Part of the reason for this is that throughout this entire period there has been a considerable amount of confidence in American industry's strengths and potentialities. This also means that policy makers have generally viewed the industrial laggardness of specific industrial technologies as ephemeral because they assumed that America's commercial S&T was insuperable in the long run. But during the past two decades this inexorable confidence has no longer matched economic reality. The other part of the reason for the federal government's lack of a firm commitment to commercialization has been the widespread political sentiment attesting to the inherent strength of the market. Commercial initiatives supported by the federal government, in other words, inevitably have dealt with the dominating political backlash: the acceptance of the assumption that the market will prevail.

The next chapter will look closely at the development of Japan's S&T policies since its devastating wartime defeat. What needs to be stressed here is that since American policy makers were generally overconfident when it came to the strength of the nation's S&T infrastructure and its industries, they too quickly dismissed Japan's potential for commercial advancement and, caught up in the fervor of the Cold War, were not inclined to envision a substantive alternative to the development of military and supporting technologies. The suggestion here is not that American industry cannot compete. Rather, the argument is that under the conditions that have existed in the United States, industry competition has become difficult mainly because the *relatively* different competition policy in Japan has made it hard for many American businesses to mount an effective and consistent challenge.

In the present period there is more concern among U.S. policy makers for commercialization issues than in the past. But this additional interest falls far short of a comprehensive and sustained commitment to growth- and people-oriented S&T policies. The majority of recent discussions and efforts to restructure the federal government's S&T policy are seriously flawed for three important reasons: (1) they assume that some monetary savings from the end of the Cold War can provide the necessary funding for the new policy that will restore America's reduced competitive edge; (2) they accept the dual-use paradigm, seeing it as an important and integral part of the solution to the nation's S&T problems; and (3) they implicitly adopt a 'trickle-down' strategy, and for this reason tend to neglect programs that would encourage economic growth while supporting workers' interests.[114]

Current S&T policy making by the federal government is not coordinated around a specific national objective intended to promote economic growth and to create good jobs. Rather, its civilian component is rudimentary and tentative. What is more, federal S&T policy still evinces the 'background radiation' left over from the big bang that initiated and then sustained the politics of the Cold War.

This is evident, for example, in the CRADAs administered by the DOD's laboratories. While the number of CRADAs administered by the DOD has grown sharply since 1989, the department's laboratories can only provide in-kind assistance to industry, such as facilities, time, and people, but not funding. Moreover, many of the DOD's CRADAs are with companies and universities that are performing defense work. The Army, which has considerably more CRADAs than the combined total for the Navy and the Air Force, estimates that over one third of its CRADAs in 1992 were with defense-engaged partners. This, for sure, deprioritizes commercial competitiveness as a post-Cold War objective. In the near-term period, about the most that can be expected from the DOD's laboratories are CRADAs that involve dual-use work.[115]

We should note that the increased concern for commercialization gleaned from U.S. government policy statements is overstated somewhat. This is evident from the frequent reference to dual-use technology, currently the S&T policy buzz word. Dual-use suggests the development of technologies that will address national security interests as they relate to both improving industry competitiveness and maintaining military preparedness. Not only does the dual-use paradigm suggest that both interests are equal, or at least nearly so, and that technology development is tantamount to successful utilization in the defense and civilian sectors, but it serves as a policy diversion by deflecting attention away from the dominant objective of military technological superiority still deeply embedded in the S&T policies of the United States.[116]

One must not lose sight of the important fact that the end of the Cold War put unprecedented pressure on the United States to take a few steps in the direction of conversion.[117] Mainly because of the crisis in S&T, which had surfaced toward the end of the Cold War period, U.S. policy makers could easily succumb at least somewhat to this pressure. These steps, however, are not the same as a sustained and growing commitment to industry development and commercialization, nor do they rule out retrogression initiated by changes in political thinking. In April 1995, for example, Congress slashed TRP funding (originally slated to exceed $600 million) by $300 million to assist in paying for expenses incurred by recent U.S. military operations abroad.[118]

The existence of some recent initiatives that use more of the national resources of the United States on commercialization activities than in the past do not change its official S&T objective. In 1992 the DOD formally proclaimed its intention to maintain S&T superiority in the aftermath of the Cold War. Although real (1992 dollars) defense spending for research, development, testing, and evaluation did decline a little between 1989 to 1991, it subsequently leveled off for a few years.[119] There is no guarantee that the small reductions in defense R&D beginning after 1995 will continue.[120]

The Clinton administration predicted the federal civilian R&D will pull ahead of military R&D by 1998 or before. In addition to the perennial problem in policy making that plans, for whatever reason, do not always translate into reality, there is an important point that needs consideration. In one accounting technique used by the Office of Management and Budget, the government includes funding for defense dual-use activities in the category for civilian R&D. In 1995 and 1996 defense dual-use technology expenditures were about $2 billion.[121] Since defense programs support most dual-use spending, it is indeed very reasonable to question the legitimacy of categorizing these expenditures as part of the government's civilian effort.

As in the past, America's current S&T policy does not demonstrate a sustained focus on economic growth while seeking a truly democratic balance between decent employment, and general improvements in the quality of life, which should include the protection of the environment. Such a focus would give significantly more attention to workers' interests and those of community and regional organizations. This would therefore require much more spending on research designed to improve education and training, which are necessary to better prepare the workforce in the manufacturing and technical areas that are important to compete with Japan in the global economy.

Notes to chapter 2

1. We use the expression science and technology policy to denote the increasing interrelationship between science and technology, rather than the narrow term technology policy. Three decades ago some analysts wisely recognized that failing to link science to technology means that 'important elements in the relationship of government to science are irretrievably lost.' Indeed, this observation is more relevant today. See E. R. Piore and R. N. Kreidler, 'Recent Developments in the Relationship of Government to Science,' in *The Annals of the Academy of Political and Social Science* January, vol. 327 (Philadelphia: American Academy of Political and Social Science, 1960), p. 13. See also Erich Bloch, 'Science and Technology: The False Dichotomy,' in *Powernomics: Economics and Strategy After the Cold War*, eds. Clyde Prestowitz, Ronald Morse, and Alan Tonelson (Lanham, MD: Madison Books, 1991), pp. 199-204.

2. See National Resources Committee, *Research -- A National Resource*, part 1, (Washington, D.C.: United States Government Printing Office [USGPO], 1938), pp. 66-67.

3. Vannevar Bush, *Science: The Endless Frontier* (Washington, D.C.: USGPO, July 1945), pp. 79-80.

4. Stephen Cohen and John Zysman, 'Diverging Trajectories: Manufacturing Innovation and American Industrial Competitiveness,' *Politics and Productivity*, pp. 39-55.

5. Bush, *Science*, pp. 7, 12.

6. Calculated from National Science Foundation (NSF), *National Patterns of R&D Resources: 1990* (NSF 90-316), pp. 47, 55.

7. Ibid.

8. Bureau of the Budget (BOB), *Budget of the United States Government for the Fiscal Year Ending June 30, 1950* (Washington, D.C.: USGPO, 1949), p. 51.

9. BOB, *Special Analysis of the United States Budget, Fiscal Year 1971* (Washington, D.C.: USGPO, 1970), p. 266.

10. NSF, *Selected Data on Federal R&D Funding by Budget Function: Fiscal Years 1990-92* (NSF 91-319), p. 26.

11. See BOB, *Special Analysis, Fiscal Year 1971*, p. 266.

12. See BOB, for example, *The Federal Budget in Brief, Fiscal Year 1959* (Washington, D.C.: USGPO, 1958), p. 44; and BOB, *The Federal Budget in Brief, Fiscal Year 1960* (Washington, D.C.: USGPO, 1959), p. 46.

13. BOB, *The Federal Budget in Brief, Fiscal Year 1958* (Washington,

D.C.: USGPO, 1957), p. 42.

14. BOB, *The Federal Budget in Brief, Fiscal Year 1964* (Washington, D.C.: USGPO, 1963).

15. BOB, *Special Analysis of the United States Budget, Fiscal Year 1967* (Washington, D.C.: USGPO, 1966), p. 122.

16. Gordon Thompson, 'The Genesis of Nuclear Power,' *The Militarization of High Technology*, ed. John Tirman (Cambridge, MA: Ballinger, 1984), pp. 66-71.

17. Ibid.; and Linda Cohen and Roger Noll, 'The Clinch River Breeder Reactor,' *The Technology Pork Barrel* (Washington, D.C.: Brookings Institution, 1991), pp. 217-257.

18. BOB, *Special Analysis, Fiscal Year 1971*, p. 266.

19. See Richard Barber Associates, Inc. *The Advanced Research Projects Agency, 1958-1974* (Alexandria, VA: Defense Technical Information Center, December 1975).

20. Ibid.

21. Congress of the United States, Office of Technology Assessment (OTA), *Commercializing High-Temperature Superconductivity* (Washington, D.C.: USGPO, June 1988), pp. 106-107. In 1972 the NSF began managing these laboratories.

22. Calculated from NSF, *Research and Development in Industry, 1983* (NSF 85-325), p. 29; and NSF, *Research and Development in Industry: 1989* (NSF 92-307), p. 45.

23. See Susan Edelman, 'The American Supersonic Transport,' in Cohen and Noll, *The Technology Pork Barrel*, pp. 97-147.

24. Seymour Melman, *Profits Without Production* (New York: Knopf, 1983), p. 83.

25. Hearings Before a Subcommittee of the Select Committee on Small Businesses, United States Senate, 88th Congress, *The Role and Effect of Technology in the Nation's Economy*, March 13, 1963, p. 405.

26. Gordon Bain, 'The Way to Supersonic Air Travel,' vol. 16, no. 7, 1963, *Sperryscope*, pp. 5-8.

27. Edelman, 'The American Supersonic Transport.'

28. Department of Defense (DOD), *500 Military Prime Contractors Listed According to the Net Value of Military Prime Contract Awards for Experimental, Developmental, Test and Research Work, Fiscal Year 1964*.

29. DOD, *100 Companies and Affiliates Listed According to Net Value of Military Prime Contract Awards, July 1950 - June 1957 and January 1955 - June 1957*.

30. DOD, *100 Companies and Affiliates Listed According to Net Value of Prime Military Contract Awards, 1 July - 30 June 1958; and DOD, 100 Companies Receiving the Largest Dollar Volume of Prime Contract Awards, Fiscal Year 1970.*

31. Edelman, 'The American Supersonic Transport,' pp. 106-107.

32. Robert Rosen and Louis Williams, 'The Rebirth of SST,' *Technology Review*, February/March 1993, pp. 22-29.

33. Terutomo Ozawa, *Japan's Technological Challenge to the West, 1950-1974* (Cambridge, MA: MIT Press, 1974), pp. 26-29.

34. Dorthy Christelow, 'Japan-U.S. Joint Manufacturing Ventures,' in *Japan's Economic Challenge*' (Washington, D.C.: USGPO, 1990), p. 212 (Study Papers Submitted to the Joint Economic Committee, Congress of the United States).

35. Ozawa, *Japan's Technological Challenge*, pp. 16-51.

36. Calculated from NSF, *International Science and Technology Data Update: 1991* (NSF 91-309), p. 5.

37. Theodore Cohen, *Remaking Japan: The American Occupation as New Deal* (New York: Free Press, 1987), pp. 353-377.

38. Akira Iriye, 'Japan's Defense Strategy,' *The Annals of the American Academy of Political and Social Science*, January 1991, pp. 41-42.

39. Ozawa, *Japan's Technological Challenge*, pp. 21, 68-69.

40. See Anthony DiFilippo, 'In Opposition: The Demise or Resurgence of Japan's Socialist Party,' paper presented at the Allied Social Science Associations Meeting, Washington, D.C., January, 7, 1995.

41. Ozawa, *Japan's Technological Challenge*, pp. 76-79.

42. Iriye, 'Japan's Defense Strategy,' p. 42.

43. BOB, *Special Analysis, Fiscal Year 1971*, p. 261.

44. Ibid., p. 256.

45. Office of Management and Budget (OMB), *Special Analysis of the United States Budget, Fiscal Year 1975* (Washington, D.C.: USGPO, 1974), p. 221.

46. OMB, *Special Analysis, Fiscal Year 1975*, p. 221; and OMB, *Special Analysis of the United States Budget, Fiscal Year 1980* (Washington, D.C. USGPO, 1979), p. 304.

47. NSF, *Federal R&D By Budget Function, Fiscal Years 1988-1990* (NSF 89-306), April 1989, p. 109.

48. OMB, *The Budget of the United States Government, Fiscal Year 1973* (Washington, D.C.: USGPO, 1972), pp. 30-31; and OMB, *Budget in Brief, Fiscal Year 1973* (Washington, D.C.: USGPO, 1972).

49. Anthony DiFilippo, *Arms Control Versus Disarmament* (Washington, D.C.: National Commission for Economic Conversion and Disarmament, May 1991, Briefing Paper).
50. OMB, *The Budget of the United States Government, Fiscal Year 1973*, pp. 30-31.
51. OMB, *The Budget of the United States Government, Fiscal Year 1973*, p. 31.
52. OTA, *Commercializing*, pp. 73-74; and Richard Thorton, 'Why the United States Needs a Maglev System' *Technology Review*, April 1991, pp. 31-42.
53. Richard Florida and David Browdy, 'The Invention That Got Away,' *Technology Review*, August/September 1991, pp. 43-54; OTA, *The Big Picture: HDTV & High-Resolution Systems* (Washington, D.C.: USGPO, June 1990), pp. 68-72; and DOD, *Key Technologies Plan* (Washington D.C.: Department of Defense, July 1992), p. 5-26. This report maintains that Japan has a 2 to 4 year lead over the United States in active matrix technology.
54. See Lee Granberg, Thomas Soller, and Harry Wise, 'Computer-Driven CRT Displays,' *Sperry Rand Engineering Review*, vol. 22, no. 3, 1969, pp. 11-17. Sperry was working on a number of advanced screen technologies, including thin screen plasma panels. See Frank Finegan, Larry Saddler, and Harry Thompson, 'Cockpit Displays for Attack Aircraft,' *Sperry Rand Engineering Review*, vol 22, no. 3, 1963, p. 36.
55. See annual reports of DOD, *100 Companies Receiving the Largest Dollar Volume of Prime Contract Awards*.
56. See DOD, *500 Contractors Receiving the Largest Dollar Volume of Military Prime Contract Awards for RDT&E*, for fiscal years 1972 and 1973.
57. OTA, *The Big Picture*, p. 91.
58. See Anthony DiFilippo, 'Why Japan is Winning the High-Tech Battle?' *Forum for Applied Research and Public Policy*, Fall 1992, pp. 15-20; OTA, *Commercializing*, pp. 45-46; and Richard Rosenbloom and Michael Cusumano, 'Technological Pioneering and Competitive Advantage: The Birth of the VCR Industry,' *California Management Review*, Summer 1987, pp. 51-76.
59. RCA losses on the VideoDisc project exceeded one half billion dollars. See OTA, *Commercializing*, pp. 46-47.
60. OTA, *The Big Picture*, p. 34.
61. See OMB, *The Budget of the United States Government, Fiscal Year*

1992 (Washington, D.C.: USGPO, 1991), pp. 51-52.

62. J. R. Powell and G. R. Danby, 'A 300-mph Magnetically Suspended Train,' *Mechanical Engineering*, November 1967, pp. 30-35.

63. For a discussion of this see DiFilippo, 'Why is Japan Winning the High-Tech Battle?' pp. 15-20.

64. See, for example, George Gilder, *Wealth and Power* (New York: Basic Books, 1981), pp. 223-224.

65. These are fully discussed in Anthony DiFilippo, *From Industry to Arms: The Political Economy of High Technology* (Westport, CT: Greenwood Press, 1990).

66. NSF, *Selected Data on R&D Funding by Budget Function: Fiscal Years 1990-92*, p. 28; and National Science Board (NSB), *Science & Engineering Indicators - 1991* (Washington, D.C.: USGPO, 1991), p. 305.

67. OMB, *Special Analysis of the United States Budget, Fiscal Year 1985* (Washington, D.C.: USGPO, 1984), pp. K-22 - K-23; and OMB, *Special Analysis: Budget of the United States Government, Fiscal Year 1986* (Washington, D.C.: USGPO, 1985), p. K-4.

68. *Report of the National Critical Technologies Panel* (Springfield, VA: National Technical Information Service, March 1991), pp. 11-14.

69. OMB, *Special Analysis: Budget of the United States Government, Fiscal Year 1987*, (Washington, D.C.: USGPO, 1986), p. K-23.

70. OMB, *Special Analysis: Budget of the United States Government, Fiscal Year 1988* (Washington, D.C.: USGPO, 1987), p. J-2.

71. Ibid., p. J-22; OMB, *Special Analysis: Budget of the United States Government, Fiscal Year_1989* (Washington, D.C.: USGPO, 1988), p. J-24; and OMB, *Special Analysis: Budget of the United States Government, Fiscal Year 1990* (Washington, D.C.: USGPO, 1989), p. J-8.

72. OMB, *Special Analysis, Fiscal Year 1980*, p. 306; and Linda Cohen and Roger Noll, 'Synthetic Fuels From Coal,' in Cohen and Noll, *The Technology Pork Barrel*, p. 288.

73. OMB, *Special Analysis, Fiscal Year 1980*, p. 297; and Linda Cohen and Roger Noll, 'The Clinch River Breeder Reactor,' in Cohen and Noll, *The Technology Pork Barrel*, pp. 236-239.

74. OMB, *Special Analysis, Fiscal Year 1980*, PP. 306-307; and William Pegram, 'The Photovoltaics Commercialization Program,' in Cohen and Noll, *The Technology Pork Barrel*, p. 350.

75. OMB, *Special Analysis: Budget of the United States Government, Fiscal Year 1981* (Washington, D.C., USGPO, 1980), p. 307; OMB, *Special Analysis: Budget of the United States Government, Fiscal Year 1982*

(Washington, D.C.: USGPO, 1981), p. 309.

76. Executive Office of the President, Office of Science and Technology Policy, *Science and Technology: A Report to the Congress* (Washington, D.C.: Office of Science and Technology Policy, March 17, 1992), p. 104.

77. See, for example, OMB, *Special Analysis, Fiscal Year 1988*, pp. J-23 - J-24.

78. See 'What's in a Name Change? For HDTV, Government Funding,' *Business Week*, December 10, 1990, p. 181.

79. See, for example, OMB, *Special Analysis, Fiscal Year 1990*, pp. J-1- J-2.

80. See National Aeronautics and Space Administration (NASA), *Commercial Use of Space: A New Economic Strength for America* (Washington, D.C.: NASA, 1989). At the same time, it is clear that NASA is still involved in DOD projects. See NASA, *NASA Aeronautics Research and Technology Program Highlights* (Washington, D.C.: NASA, 1989).

81. OMB, *Special Analysis, Fiscal Year 1989*, pp. J-14 - J-15; and OMB, *Special Analysis, Fiscal Year 1990*, p. J-2.

82. Allan Bromley, *U.S. Technology Policy* (Office of Science and Technology Policy, September 26, 1990), p. 6.

83. On federal support for high-temperature superconductivity see OTA, *Commercializing*.

84. See Anthony DiFilippo, *How the Military-Serving Firm Differs From the Rest*, (Washington, D.C.: National Commission for Economic Conversion and Disarmament, March 1991, Briefing Paper); OTA, *Commercializing*, p. 98; and OTA, *Holding the Edge* (Washington, D.C.: Office of Technology Assessment, April 1989, Summary), p. 29.

85. Quoted in 'Will Uncle Sam Be Dragged Kicking and Screaming Into the Lab?' *Business Week*, July 15,1991, p. 129.

86. See Bromley, *U.S. Technology Policy*; Office of Science and Technology Policy, *Science and Technology: A Report to Congress*, p. 94; and B. R. Inman and Daniel Burton, 'Technology and Competitiveness: The New Policy Frontier,' *Foreign Affairs*, Spring 1990, pp. 116, 124.

87. Cecil Uyehara, 'Appraising Japanese Science and Technology,' in *Japan's Economic Challenge* (Washington, D.C. USGPO, 1990), pp. 295-296 (Study Paper Submitted to the Joint Economic Committee, Congress of the United States).

88. DOD, *Critical Technologies Plan* (Washington, D.C.: Department of Defense, May 1, 1991); and DOD, *Key Technologies Plan*, July 1992).

89. Shintaro Ishihara, *The Japan That Can Say No* (New York: Simon &

Schuster, 1991), pp. 20-22.

90. DOD, *Critical Technologies Plan*, p. 1-115.

91. See *Report of the National Critical Technologies Panel.*

92. See Lewis Branscomb, 'Dual-Use Technology: Optimizing Economic Security Interests Through National Technology Policy,' (Cambridge, MA: Science, Technology and Public Policy Program, John F. Kennedy School of Government, Discussion Paper 89-08); Ashton Carter, 'Spin-Offs and Spin-Ons: Civilian and Military R&D,' in *Science & Security: Technology Advances & the Arms Control Agenda*, eds. Eric Arnett, Elizabeth Kirk, and W. Thomas Wander (Washington, D.C.: American Association for the Advancement of Science, 1990), pp. 163-177; John Alic, Lewis Branscomb, Harvey Brooks, Ashton Carter, and Gerald Epstein, *Beyond Spinoff: Military and Commercial Technologies in a Changing World* (Cambridge, MA: Harvard Business School Press, 1992); and William Gregory, *The Price of Peace: The Future of the Defense Industry and High Technology in a Post-Cold War World* (New York: Lexington, 1993).

93. See Murray Weidenbaum, *Small Wars, Big Defense* (New York: Oxford University Press, 1992), p. 99; and Murray Weidenbaum, 'Filling in the Hollowed-Out Corporation: The Competitive Status of U.S. Manufacturing,' *Business Economics*, January 1990, p. 21.

94. NSF, *The Science and Technology Resources of Japan: A Comparison with the United States* (NSF 88-318), p. 56.

95. See National Research Council, *Expanding Access to Precompetitive Research in the United States and Japan: Biotechnology and Optoelectronics* (Washington, D.C.: National Academy Press, 1990); and National Research Council, *Learning the R&D System: Industrial R&D in Japan and the United States* (Washington, D.C.: National Academy Press, 1990).

96. OMB, *Budget of the United States Government, Fiscal Year 1992*, p. 47.

97. See 'Here Comes the Thinking Car,' *Business Week*, May 25, 1992, pp. 84-87; and OMB, *Budget of the United States Government, Fiscal Year 1992*, pp. 51-52.

98. Bill Clinton, *Technology: The Engine of Economic Growth* (Little Rock, AR, September 21, 1992), p. 14.

99. President William Clinton and Vice President Albert Gore, *Technology for America's Economic Growth: A New Direction to Build Economic Strength* (Washington, D.C.: USGPO, February 22, 1993), pp. 8-9.

100. OMB, *Budget of the United States Government, Fiscal Year 1995* (Washington, D.C.: USGPO, 1994), pp. 122-123.

101. See ARPA, *Program Information Package for Defense Technology Conversion, Reinvestment, and Transition Assistance*, March 10, 1993; and 'Technology Conversion Spearheaded by ARPA,' *National Defense*, May/June 1993, pp. 20-21.

102. OMB, *Budget of the United States Government, Fiscal Year 1995*, p. 123.

103. ARPA, *Program Information Package for Technology Reinvestment Project, Focused Competition*, April 1994.

104. Jacques Gansler, 'Restructuring the Defense Industrial Base,' *Issues in Science and Technology*, Spring 1992, p. 53.

105. Calculated from *OMB, Budget of the United States Government, Fiscal Year 1995*, pp. 118-122.

106. On the Critical Technologies Institute, see 'A Think Tank for, Er, 'Competitiveness,'' *Business Week*, April 20, 1992, p. 90.

107. OMB, *Budget of the United States Government, Fiscal Year 1992*, p. 57; and OTA, *Competing Economies: America, Europe, and the Pacific Rim* (Washington, D.C. USGPO, October 1991), pp. 62-63.

108. Quoted in 'A Think Tank for, Er, 'Competitiveness,'' *Business Week*, April 20, 1992, p. 90.

109. OMB, *Budget of the United States Government, Fiscal Year 1995*, p. 117; and OMB, *Budget of the United States Government, Fiscal Year 1996* (Washington, D.C.: USGPO, 1995).

110. OMB, *Budget of the United States Government, Fiscal Year 1992*, p. 57.

111. Of the eight agencies supporting this effort in 1992, DARPA, NASA, and Energy were slated to receive more than 60 percent of the total. See OMB, *Budget of the United States Government, Fiscal Year 1992*, p. 49.

112. See OTA, *Contributions of DOE Weapons Labs and NIST to Semiconductor Technology* (Washington, D.C.: USGPO, September 1993), pp. 3-7; and OTA, *Defense Conversion: Redirecting R&D* (Washington, D.C.: USGPO, May 1993), pp. 103-114.

113. OMB, *Budget of the United States Government, Fiscal Year 1995*, p. 118; and OMB, *Budget of the United States Government, Fiscal Year 1996*.

114. See Lewis Branscomb, 'Does America Need a Technology Policy?' *Harvard Business Review*, May-April 1992, pp. 24-31; and Alic et al., *Beyond Spinoff*.

115. OTA, *Defense Conversion*, pp. 152-158.

116. For example, see OMB, *Budget of the United States Government, Fiscal Year 1995*, p. 223.

117. See OTA, *After the Cold War: Living With Lower Defense Spending* (Washington, D.C.: USGPO, February 1992).

118. Technology Reinvestment Project, *TRP Update* (Technology Reinvestment Project: Arlington, Virginia, April 26, 1995); and Technology Reinvestment Project, *Joint Program Solicitation* (Technology Reinvestment Project: Arlington, Virginia, May 12, 1995).

119. Director of Defense Research and Engineering, *Defense Science and Technology Strategy* (Washington, D.C.: Department of Defense, July 1992), p. I-2; and OTA, *Defense Conversion*, pp. 148-149.

120. OMB, *Budget of the United States Government, Fiscal Year 1995*, p. 116.

121. Ibid., pp. 118-119, 122; and OMB, *Budget of the United States Government, Fiscal Year 1996.* In fiscal year 1996, the civilian proportion of the U.S. government's R&D budget came to 51 percent when dual-use spending was included and to 48 percent when it was not.

3 Behind economic growth: Japanese science and technology policy from inception to maturity

During the decades after World War II Japan continued to develop, eventually, during the 1980s, becoming the second leading economic power in the world. Although it is clear that a consistent commitment to capital expenditures that have been fueled by a high national savings rate, when combined with a strong work ethic, have been instrumental to Japanese post-war economic growth, this chapter will focus on Japan's industrial strength and the significance of it developing and effectively managing technology and, subsequently, science. Since 1988 Japan has devoted a larger proportion of its gross domestic product to research and development than has the United States, even after including American military expenditures.

This chapter then rests its argument on the very reasonable assumption that the development of science and technology (S&T) have propelled Japan's manufacturing industries and have been very important in it becoming a leading economic power. Further proof of Japan's position as an established global economic power has been its increased efforts to improve the quality of its science and to integrate this with its proven technological skills. In order to understand fully the development of Japanese S&T policies over the past several decades and how they have powered industrial advancements, however, it is necessary to look briefly at major developments prior to World War II.

The pre-war period

In the early 1930s the leaders of the *zaibatsu* (very concentrated industry groups) had come to recognize that their economic interests were not antithetical to those who commanded Japan's military systems. Whereas some disagreements had previously existed between the *zaibatsu* and the military, by 1931 their sometimes divergent interests could no longer be discerned. This convergence of interests was specifically the result of the military's success over the Chinese in Manchuria; it was because of this that the *zaibatsu* leaders began to appreciate the economic advantages of turning more aggressively to war production.[1]

Because of this definitive move to war production, major changes occurred in Japanese industry. One important change was that the Japanese strove for increased technological independence by relying less on foreign sources.[2] Another significant change was that more emphasis was placed on the technical development of the machine tool, automobile, aircraft and other 'heavy' industries. This meant that 'light' industries such as textiles which had developed rapidly after World War I became of secondary importance to the Japanese.[3] Japan's machine tool industry had remained relatively undeveloped during the interwar years and there was a noticeable difference in quality between domestic- and foreign-built tools. Without a strong domestic machine tool industry, war production would be a futile effort. Thus efforts to develop the machine tool and other still underdeveloped durable goods industries was imperative if war production was to proceed.

The mobilization of Japan for war became politically visible in 1931 with the passage of the Law of Control of Key Industries. This law had the effect of strengthening the economic power of the *zaibatsu* and at the same time it mandated the regulation of industry for war production. Subsequent legislation during the 1930s such as the War Industry Mobilization Law crystallized Japan's intention to direct industry to military preparedness.[4]

By 1938 the political and economic determination to improve the efficiency of military production manifested itself in the National General Mobilization Law. More so than in the past, this law was intended to increase control over Japan's capital and labor resources for the explicit purpose of national defense. Significantly, Japan not only had fully mobilized industry for national defense by the late 1930s, but its S&T resources had also been largely appropriated to support this objective. Also in 1938 a Science Council directed by the Prime Minister's office was formed, as was the Science Encouragement and Investigation Committee headed by the Ministry of Education; both organizations had military objectives. The next year the Cabinet Planning Board created for itself a Science Department expressly to coordinate science and research activities for the purpose of war. The concern in Japan with developing its military capabilities was again reflected in 1941 with the creation of an office of technology. Military personnel labored diligently to control this office which was established as a coordinating agency both for the efforts of scientists and engineers in the private sector and generally to better manage S&T work. Military personnel were successful in centralizing work to support S&T for aviation development. After gaining leadership positions in many of the branches of the technology office, military personnel began restricting work performed by scientists and engineers.[5]

By the late 1930s the increased emphasis on war production and the attendant issue of appropriating S&T resources was evident as the government forced companies to put critical civilian projects aside to accommodate military work. Although preparing for war consumed the bulk of manufacturing and supporting S&T efforts, some important consumer work (e.g., in consumer electronics) still existed in Japan at this time. With Tokyo selected as the site of the 1940 Olympics, one major manufacturing concern was to establish television broadcasting technologies to air the games. This work, however, could not be completed because of World War II.[6]

How this affected Matsushita is an interesting aside to this issue. In 1938 Matsushita established a research institute in Tokyo. A major research focus of the institute at the time was on developing TV broadcasting technologies for the Olympic games. But Matsushita executives also had their eyes trained on R&D work supported by the military, specifically developing radio apparatuses for Japan's army and navy. Once World War II began Matsushita had to abandoned civilian TV work; it was required to make many different military products, such as machine guns, boat engines, shells, and -- more in line with the company's emerging electronics talents -- radios for planes and land vehicles, and vacuum tubes. In addition to getting technical assistance from the military, Matsushita also received R&D support from Toshiba, which at the time was responsible for transferring military electronics technologies to other companies.[7]

Occupying Japan: converting S&T

Japan's defeat in World War II not only led to a protracted occupation by the United States but, importantly and unexpectedly, this period was also a new beginning for the development of industrial S&T. Because of directives from the Supreme Commander for Allied Powers (SCAP) during the occupation period, Japan's industrial and technology bases were effectively transformed from fulfilling military objectives to meeting civilian ones -- at least for several years.

Three weeks after the acceptance by the United States of Japan's surrender, SCAP issued a directive calling for the demilitarization and disarmament of Japan. Reflecting an awareness of the accumulated power of the *zaibatsu* to wage war, the SCAP directive dated September 6, 1945 called for the immediate end of military production in Japan. Two weeks after this another SCAP directive provided the complete details for converting Japan's factories from military to civilian production. The SCAP directive dated September 22, 1945 had far-reaching effects on Japan's S&T. This directive imposed major demands on Japan's S&T institutions such as monthly reports on research and science activities, and the description of all work done after 1940; these institutions were also put on notice that the inspection of facilities could occur at any time. By mid November another SCAP directive prohibited even commercial aviation production and research after the end of December 1945. Thus SCAP directives were aimed squarely at the *zaibatsu* because of their involvement in military production since the early 1930s. It is important to understand that *zaibatsu* linkages to war manufacturing had not only been direct but they were also connected to different military organizations. For example, the Mitsui *zaibatsu* attended to the needs of the army, while the Mitsubishi *zaibatsu* met the demands of the navy.[8]

The determination of the United States to dismantle completely Japan's industrial and S&T war machine became evident again seven months later. In late June 1946 the U.S. Department of State issued a Draft Treaty on the Disarmament and Demilitarization of Japan. Endorsed also by China, the Soviet Union, and Great Britain, this document noted that the removal of Japan's war capability 'will be enforced as long as the peace and security of the world may require it.'[9] This treaty was issued in the wake of the U.S.-approved peace constitution for Japan which was presented by the Japanese government to SCAP on April 22, 1946.[10] Thus the occupation marked the beginning of a very paternalistic relationship between Japan and the United States.

Interestingly, paternalism had long been a part of Japan's domestic development. But this 'endemic' paternalism reflected the internal dynamics of a developing capitalist economy. A good example of this paternalism can be seen in Japan's textile industry. Much earlier in the twentieth century 'friendly societies' emerged in Japan's textile industry. These societies spread through the textile industry and, although less common, were still in existence in the 1950s. The paternalism attached to the 'friendly societies' took the form of employers providing benefits in-kind to what were often young, low-paid female workers who came from rural areas to work in the mills for a few years before marriage. With SCAP providing leadership to the Japanese economy after the war, the labor movement grew and the transition to modern labor management began to appear in the textile industry.[11]

Although many things had changed in Japan, by 1945 paternalism was still visible. Now, however, it was business and government that had to succumb to a higher power, and unlike in the past, this 'new' paternalism was exogenously imposed on Japan. This paternalism covered all aspects of Japanese life: the economy, politics, trade, and S&T. As for the parts of the Japanese population afflicted by widespread food shortages, they were indifferent to the enveloping paternalism. As one Japanese newspaper put it: 'If we can be feed we don't care who has political power.'[12]

While paternalism characterized the overall relationship between the United States and Japan, the continuing and successful efforts to remove military production and R&D from Japan during the early post-war period became a Pandora's box for the United States by the time of the Korean War. This, however, was only the first of several of these boxes that appeared throughout the post-war years. The problems that emerged unexpectedly touched on critical military, S&T, political, and trade issues.

The paradoxes of paternalism

As U.S. officials were busy preparing America's national S&T policies to attend to the needs of the nation's newly assumed position as a military superpower, they had purposely assigned Japan's S&T to civilian activities by 1950. Not only was Japan seen as a security risk to the world, but the extensive monitoring of and constraints on its S&T activities by the occupation forces meant that the United States had exercised complete circumspection by preventing any technological rekindling of military nationalism. By focusing on commercial objectives and with S&T as a support to industry, Japan's only practical option was to concentrate on civilian production. This, however, did not last too long.

The new Japanese constitution was just slightly over four years old when the Korean War broke out; this conflict bolstered the still-depressed Japanese economy in two ways. On the one hand, the Japanese economy improved because of the increased demand for civilian goods, since U.S. industries had more decidedly turned to military production. Older industries such as iron and steel, shipbuilding, and electric power had been rebuilt in the early 1950s, while newer ones such as consumer electronics, transportation equipment, and petrochemicals emerged and further stimulated economic growth.[13] It was during the early 1950s that Japan's consumer electronics industrialists began to show signs of market foresight. In 1950 Sony constructed the initial sound magnetic recorder; three years later it built a television camera, and by 1955 it had introduced a stereo tape recorder.[14]

On the other hand, America's involvement in the Korean War created a demand for weapons sufficient enough to generate arms production not only in the United States, but in Japan also. By March 1951 the Law for Promoting Industrial Rationalization had been passed, thereby formally setting in motion the legalized and U.S.-approved remilitarization of Japan. Understanding that Japan's machine tool sector was not at the time highly advanced, and because of its central importance to weapons production, a group whose responsibility was to investigate its problems, arrived in Japan from the United States in February 1952 to assist in the development of the industry. Moreover, rearming Japan meant that its S&T resources also had to remobilized for war. This, indeed, did occur during the Korean War. By March 1952 SCAP had approved aircraft work in Japan with some restrictions. In addition to aircraft research, SCAP proscriptions on research activities in other areas such as radar, and atomic energy were also dissolved in Japan during the Korean conflict, with work refocused on military objectives. The U.S. military also made use of Japanese scientists and other technical personnel during the Korean War.[15]

Particularly interesting is the ending of the ban on Japan's aircraft industry during the Korean War. At the beginning of Word War II Japan's military aircraft power compared favorably to that of the United States, despite its much smaller economy.[16] Although Japan's aircraft and production facilities had either been destroyed in battle or put out of business by SCAP directives, neither of the two could abolish the human know-how necessary for design, manufacture, and repair. During the Korean conflict, the United States used Japanese aircraft skills to repair and maintenance aircraft.[17] Even though the occupation of Japan had separated the Mitsubishi *zaibatsu*, which was comprised of many companies and which was responsible for manufacturing the Zero Fighter planes during World War II, the United States relied on Mitsubishi Heavy Industries during the Korean conflict to construct the F-86 fighter.[18]

Japan's peace constitution therefore did not apply to the needs of the United States, which required access to Japanese technology and manufacturing capabilities. SCAP-imposed prohibitions on Japan that converted its economic and S&T bases immediately after World War II were removed during the Korean conflict -- but only because of the military requirements of the United States. Nonexistent prior to the Korean War, U.S. procurement from Japan, the majority of which was for military purposes, increased markedly from $149 million in 1950 to $592 million by the end of 1951. Total U.S. procurement from Japan between 1951 and 1953 was $2,374 million, and an estimated 60 to 70 percent of its exports at the time were stimulated by war spending.[19] Because Japan would not be an aggressor in the Korean conflict, American policy makers read the Japanese peace constitution with one blind eye and so were able to disregard its strictures stipulating the renunciation of war. The Japanese, however, ultimately justified partial rearmament somewhat differently. They saw it as a way to stimulate their depressed economy.

By the time the San Francisco Peace Treaty became operative in April 1952 (it had been signed the previous year) marking the end of the U.S. occupation, Japan had turned its S&T toward military production to a limited extent. Even though military activities were reduced after the Korean War ended in the summer of 1953, Japan was not fully disarmed. Too much had happened since 1950 for this to occur. Within two months after the beginning of the Korean War, SCAP approved the formation of Japan's National Police Reserve; by 1954 this evolved into the Self Defense Forces. The signing of the San Francisco Peace Treaty occurred on the exact day as the first U.S.-Japan Security Treaty which authorized a permanent U.S. military presence in Japan. Thus the combination of the enduring U.S. military presence and the existence of the Self Defense Forces meant that S&T for military support purposes was not to be excised completely from the Japanese economy.[20]

This was true well after the dust from the Korean War had settled.

America's new global military responsibility led to a protracted military relationship between these two nations. Japan serviced the U.S. military through its emerging manufacturing and technology bases. In the late 1950s Nippon Electric Company was let a contract to build a long distance transmission apparatus called Tropospheric Communications. Relying on U.S. technology, Nippon successfully completed the project and even managed to make significant contributions on its own to the system. So satisfied were U.S. military leaders with the outcome of this project that they issued a contract for the building of another system from the Japanese. Although this and other types of military manufacturing and technology support continued, the argument that the U.S. military presence in Japan encouraged technology transference and licensing agreements which significantly supported industrial and technology development there is exaggerated.[21]

While it is true that by 1955 Japan had surpassed its production peak reached before World War II at least partly because of the economic stimulus created by the Korean War,[22] it is a completely different point to allege that its mainly civilian S&T efforts were fundamentally supported and encouraged by the U.S. military presence there. That the Japanese manufactured aircraft for the U.S. military, for example, hardly supports this transference argument, since this industry remained one of the least developed in Japan.

Moreover, Japanese electronics systems were not compatible with U.S. military specifications. In 1959 Mitsubishi Heavy Industries won a contract to produce the F-104 Starfighter with Lockheed. After an agreement had been worked out between the United States and Japan, Mitsubishi signed a contract in the spring of 1961 with Japan's Defense Agency to manufacture two hundred of these planes in the C-1 program which ended in 1965. While Japanese firms manufactured the majority of the airframes for the C-1 program and increasingly became more independent in manufacturing the engines for the F-104J(apan), this was not the case for the electronics systems used in these planes. Even though Japanese firms had developed expertise in consumer electronics, it was of no use to the planes in the C-1 program. The electronic systems for these planes came from the United States. Japanese industry officials maintained that the differences between military and consumer electronics were too great in configuration and specification to be interchanged successfully.[23]

The continuation of military production and the support of S&T for this work notwithstanding, the Japanese concerned themselves mainly with developing a peace economy. It is, however, important to emphasize again that the complete disarmament of Japan that took effect after World War II, but which was reversed during the Korean War, was never again to reappear there. Much of the reason for this was because of the change in the policy of the United States toward Japan. The 1952 Security Treaty between these two nations stipulated that Japan should take on a more active part in its defense, but always avoiding the danger of becoming a military aggressor. Indeed, when vice president Nixon visited Japan in 1953 he acknowledged in a speech that the United States had been wrong to impose disarmament on the nation.[24] Because it did not violate the Mutual Defense Assistance Agreement of 1954 between the United States and Japan, the Technical Agreement that began in 1956 allowed for the 'exchange of patent rights and technical knowledge for defense purposes.'[25]

Establishing an industrial technology base

The SCAP-sanctioned constitution which prohibited Japan from becoming a military threat reflected the widespread view of the people. The majority of Japanese people had their fill of war and were only anticipating the economic growth that would engender improved living conditions. As a result the Japanese were exceptionally compliant with SCAP leadership.[26]

To a large extent, the Japanese directed their national S&T efforts at agriculture during the 1950s. This is understandable, given the widespread hunger problem that they had to deal with after World War II. The Japanese were also very concerned with health problems at this time and responded by organizing the National Institute of Health and by promoting the manufacture of pharmaceutical products.[27]

Japan, however, did not overlook the need to develop a broad manufacturing base. Unlike U.S. policy makers, the Japanese had no reservations about using government to generate economic growth. Thus the Japanese government embraced an interventionist approach that focused on rebuilding the commercial economy. Beginning in 1950 with the passage of the Law Concerning Foreign Investment the Japanese government aggressively sought to promote industrial growth. Because this law encouraged the importation of foreign technology, especially from the United States, it facilitated Japan's efforts to technologize industry.

Between 1956 and 1959 Japanese payments for imported technology know-how (licenses, patents, etc.) grew by more that 85 percent. From 1956 to 1965 these payments increased by almost 400 percent. The Japanese not only imported technology but they were bringing it into the country very soon after being developed by foreign sources. Moreover, between 1945 and 1954 over 90 percent of the imported technology know-how was put to use within one year, notably different from what existed in Japan before World War II. After 1955 this practice reached 100 percent, indicating clearly Japan's determination to bring technology swiftly into industrial use.[28]

For some in Japan, the 1950s was the decade of the 'three status symbols'-- the television, the electric washing machine, and the electric refrigerator. The 1950s also brought other technology-based developments into some Japanese households such as the vacuum cleaner, the electric rice cooker, the transistor radio, and the initial broadcasting of FM radio by the government-owned NHK.[29] But it was foreign technology that spawned much of the industrial development in Japan during the 1950s and early 1960s. For example, between 1957 and 1961 two fifths of the new goods developed by Japanese producers could be traced to foreign technology.[30] The upshot is that these activities helped improve the quality of life for some of the Japanese people.

While Japan was clearly dependent on foreign technology during the early post-war years, one should not assume that it simply imported these technologies and did not develop them any further. True, Japan was technologically behind the United States and Western Europe at this time, but the nation's evolving technological competence should not be ignored. For example, a number of Japanese firms were performing pioneering research on video tape recorders in the 1950s. Toshiba, Sony, Matsushita, and the Japan Victor Corporation were all involved in the early innovative research on video tape recorders. Decades later Japanese producers came to dominate the world VCR market after U.S. companies discontinued research this area.[31]

A common tactic used by Japanese researchers was to examine thoroughly imported technologies. When researchers at the government-owned Japan National Broadcasting Corporation imported a very expensive video tape recorder built by the U.S. firm Ampex in 1958, they brought in technical people from several Japanese companies to study the device. Because the government did not want to see the economy flooded with these foreign-built recorders, the Ministry of International Trade and Industry (MITI), in addition to providing technological support, made available financial assistance to companies interested in this work.[32] In short, Japanese researchers during this early period often used a sophisticated process of 'reverse engineering' on foreign technology, and with government assistance they worked to develop the technologies they imported from other nations.

However, government support of industry in Japan did not originate in the early post-war period. On the contrary, it has had a long history there. Nor did the desire to become an exporting nation originate during this time. Both are part of the economic history of Japan that actually predate World War I.

In the second half of the nineteenth century the Meiji state attempted to stimulate production so that Japan would become an export-driven economy. In 1870 the government established the Ministry of Engineering for the purpose of supporting the development of what were at the time considered to be key sectors of the emerging capitalist economy -- railroads, communications, and mining. By the turn of the century arms production became more important to the Meiji government. As a result, government began to support the development of heavy industries, and it actually purchased enterprises for this purpose. Indeed, the desire to create an industrial policy that depended on a business-government partnership can be traced back to the middle of the 1880s.[33]

It was after World War II, nonetheless, that the Japanese government demonstrated a sustained commitment to industry development. In addition to the Law Concerning Foreign Investment, the Diet passed other important legislation during the 1950s. Recognizing that it was imperative to strengthen the machine tool industry if Japan was truly to become an industrial power, the government passed the Extraordinary Measures Law for the Promotion of Specified Machinery Industries in 1956. This law permitted MITI to guide or 'target' the machine tool industry. At the same time, by emphasizing the development of domestic automobile production, the machine tool industry would grow stronger.[34] For consumer electronics, the government passed the Extraordinary Law for the Promotion of the Electronics Industry.

The Japanese government backed up these laws with yen. Approximately 10 percent of the government budget was annually set aside for industrial development. Realizing the connection between infrastructure and industrial growth, the Japanese government also earmarked funds for roads, harbors, and related projects. The government levied high taxes on the Japanese people to pay for these initiatives.[35] Despite the ongoing efforts of government to promote economic growth, the Japanese standard of living remained relatively low. Thus through the really tough times the Japanese people accepted the thriftiness imposed on them by government; it was this sacrificing that helped the economy grow.

During the 1950s the Japanese government also expanded its organizational capacity to develop industrial research. In its continuing effort to augment the coordination of the policies taking shape in Japan since 1948 when the Agency of Industrial Science and Technology was started, the government began the Science and Technology Agency in 1956.[36] In February 1959 the Japanese government authorized the creation of the Council for Science and Technology. As part of the prime minister's office, the council became responsible for the overall coordination of the nation's S&T policy.[37]

For the most part, Japan sought opportunities that would generate industrial growth. As we have seen, the Japanese took advantage of the economic opportunities that became available from the Korean War. Willing to accept American security and the position of a subordinate in the U.S.-Japan alliance, Japanese policy makers reasoned that civilian industry development supported by an evolving S&T base would provide the nation with the quickest route to accomplish the goal of strong economic growth. While keeping a toe in military activities, Japanese policy makers could still keep both feet planted firmly on civilian soil, and accordingly were fully committed to commercial development. This, they understood, required technologies that they did not yet possess. By continuously building up their ability to coordinate S&T activities, government could assist industry in exploiting imported technologies.

For the price of accepting U.S. political and economic leadership, Japan received military security. Whether this was needed or not is unimportant, since the Japanese believed that they did need this security. Determined to maintain global equilibrium, U.S. policy makers were eager to supply it. Having already experienced the manifold constraints imposed by the occupation, Japan's acceptance of U.S. security deepened the control America had over the nation. This paternalism notwithstanding, Japan was largely free to pursue the civilian objective of developing industry as quickly as possible. The Japanese realized this could only be done by incorporating as much S&T into their efforts as possible.

Although much of the growth during the early post-war years can be accounted for by imported technology, there was increased attention being given to science. Beginning about 1950, Japanese achievements in science were discernible in areas such as medicine and chemistry.[38] By 1955 the

Japanese government instituted legislation that authorized scientific research on nonmilitary atomic energy.[39] In short, during the early 1950s a scientific wave was forming in Japan, though it would not be of much significance for at least another decade.

For the United States, there was the hidden costs of paternalism. Having the most sophisticated S&T system at its disposal, policy makers, industry executives and other observers in the United States gave no thought during the 1950s to the building of technology-based manufacturing in Japan and the beginning of independent science initiatives. What mattered to U.S. policy makers was the continued development of Japan's economy. If needed, Japanese industry could serve the U.S. military swiftly, as it did during the Korean War. In the meantime, a healthy Japanese economy with an expanding S&T base could only be viewed auspiciously by U.S. policy makers. But U.S. paternalism meant more than this. Japan would remain politically and economically dependent on the United States, fully protected by America's military and relying heavily on its technology.

Militarily, this meant that Japan would get a 'free ride,' at least for awhile, since it did not have to spend excessively on defense; but it would also get an economy fare for technology know-how, because its developmental costs were lowered by importing much of this from the United States. The many technology transfers to Japan occurred because U.S. corporations used this as a strategy to increase profits. American corporations were complacent with this arrangement, for they felt that Japanese companies were too technologically underdeveloped to be a major threat to them.[40] In the short run U.S. industry executives were right, but they were quite mistaken in the long run.

Building momentum

By recognizing the importance of well qualified R&D personnel, Japan began to increase its efforts to build a strong civilian industrial sector during the 1950s. Beginning about 1950 a growing number of Japanese scientists and other technical personnel, including some from industry, traveled to the United States to study, most of which went for the purpose of broadening their knowledge of technology.[41] At the same time, there was a dramatic increase in the number of researchers in Japan.

Between 1953 and 1955 the total number of researchers in Japan increased by 58 percent and continued to increase substantially throughout the rest of the decade. Throughout the 1950s by far the majority of researchers in Japan were located in universities. During the 1950s universities were also spending more on research than companies and research institutes, many of which were public. Because Japan was just beginning its ascent from the economic devastation precipitated by World War II, the private sector could not afford to spend heavily on research. For example, in 1953 Japanese companies expended only about half of what universities devoted to research; spending by research institutes was even lower than corporate spending. By 1960 things had changed dramatically. Total research spending in Japan had markedly increased, as had the number of researchers. Research spending by companies made up more than half of the total, and there was a big increase in the number of researchers employed by companies. Research expenses by research institutes had also noticeably increased by 1960. While most researchers were still found in universities, both research spending and the number of researchers in Japan had spread to other sectors.[42]

Thus the 1950-1960 period was an important one in the development of Japanese S&T. Since companies did not have sufficient capital to make the R&D investments needed to fulfill the national objective of commercial advancement, universities and, to a lesser extent, public institutions provided supporting efforts. Little changed from three years earlier, in 1956 company research expenditures in Japan came to only 30 percent of the national total, with the rest of these investments coming from universities and research institutions. Later, because of the strengthened Japanese economy, companies had more capital to invest in research; by 1960 the corporate contribution had grown to 56 percent of the total.

Toward developing an indigenous S&T base

As alluded to above, the Japanese had much more (foreign) technology than science in the early 1960s. Nonetheless, with a determination similar to that demonstrated when they decided to import and use foreign technology, the Japanese began to promote 'independent' science in the 1960s. Once this decision was made, the Japanese slowly, yet steadily, worked to fulfill this objective. Although they took major strides to develop indigenous science in the 1960s, the Japanese understood its importance well before this time. For example, as early as 1951 Japan enacted tax laws for the purpose of scientific development. These laws exempted 'scientific research corporations' from paying various forms of research taxes.[43]

The Japanese plan for commercial development was a broad-based one. It contained the sober realization of what Japan should do at the time as well as what it could do in the future. Thus there began a national campaign to develop independent science and engineering and to integrate them with existing and new technologies; moreover, infrastructure expansion and capital investments would continue to support industry modernization.

Long considered to be the most educated society in the world, Japan began to take science and engineering training even more seriously in the 1960s. Because of encouragement from government, many more young people began seeking careers in science and engineering during the 1960s. In 1960 the Japanese government implemented a tax measure which permitted companies to take tax deductions for money donated 'for urgently needed education or science promotion expenditures.'[44] This policy paid off. Researchers in Japanese corporations increased from 42.9 thousand in 1960 to 61 thousand in 1965; by 1969 companies employed over 85,000 researchers. Thus during the 1960s the number of researchers nearly doubled in Japanese firms.

Quite obviously concerned about infrastructure development, the Japanese government supported the construction of highways, railroads, and improved telecommunications systems.[45] Also during the 1960s Japan National Railways, at the time a public institution, began exploring ways to develop superfast trains, which later led to the interest in building magnetically levitated trains.[46] In 1964 the government owned NHK Corporation started studying the prospect of high definition television, and in six years introduced a structured development program.[47]

To further support S&T advancement, the Japanese government instituted laws that encouraged corporations to take bigger steps to perform research work. In 1961 the government passed a tax law to assist mining and manufacturing companies. This law also allowed for reduced taxes on cooperative industrial work involving experimental research.[48] Thus an antitrust mentality was excised from government's view of how the private sector should operate. This antitrust excision, accordingly, strengthened the *keiretsu* (existing industry associations).[49] Not only was the cooperative approach to conducting industrial R&D seen as a critically important way by which to foster innovation, it also reduced duplication.

In 1967 the Japanese government enacted the Tax Deduction on Experimental and Research Expense Increments Measure. This law permitted companies to take a 20 percent deduction on the growth in research expenses from the highest point reached in the past.[50] Government support of industrial innovation was further evidenced by the creation of the Research and Development Corporation in the 1960s.[51]

Government efforts to promote the expansion of the nation's research efforts, especially the measures implemented to generate additional private sector spending, were largely successful. Compared to the relatively judicious spending that took place in the 1950s, company research expenditures increased dramatically during the 1960s. (See Table 3.1.) The fivefold increase in research spending by Japanese companies indicated that they were preparing themselves for the challenges of the 1970s and beyond. (By 1980 Japanese companies were financing three fourths of Japan's total R&D spending; by 1990, this fraction increased to over four fifths.) Although spending by companies accounted for nearly 60 percent of the nation's total research by 1969, research institutes were also spending increasing amounts for this purpose. Still the proportion of total research attributable to these institutes slowly declined through the decade as Japanese corporations continued aggressively to pursue research work.

The main point here is that Japanese industries were supported by government in moving the nation along the path of substantial economic growth during the 1960s.[52] Through creative tax laws, businesses could acquire the advantages of developing an innovative S&T base. Through government spending on R&D to encourage risk and, significantly, to foster the formation of a long-term perspective for S& T development, businesses continued to innovate but simultaneously felt secure in what they were doing -- an 'uncertainty principle' both created and appeased by government action. As this was taking place research expenses by universities increased phenomenally throughout the decade. In 1960 university research expenses were $400 million (1982 dollars); by 1969 they exceeded $1,812 million. Similar to what occurred in public institutions, the number of university researchers markedly grew during the 1960s, more than doubling in size from the beginning to the end of the decade.

The most significant cooperative effort to emerge in the 1960s, and one that had profound and lasting effects on the civilian sector, was the Large Scale Project. Begun in 1966 and under the general direction of the Ministry of International Trade and Industry (MITI), the Large Scale Project, bearing the official name of the National Research and Development Program, forged a model cooperative relationship between industry, government and universities. With MITI assuming the capital risks, S&T projects that perhaps would have never emerged in Japan did so,

Table 3.1 Real corporate research and development spending in Japan, 1953-1970

Year	1982 Dollars (in millions)
1953	$91.53
1954	135.14
1955	145.80
1956	154.34
1957	190.32
1958	255.25
1959	298.22
1960	863.61
1961	1,101.81
1962	1,433.92
1963	1,529.76
1964	1,758.51
1965	1,999.03
1966	1,989.04
1967	2,248.86
1968	2,810.25
1969	3,541.68
1970	4,183.70

Sources: Calculated from Statistics Bureau, Management and Coordination Agency, Historical Statistics of Japan (Tokyo, n.d.); United Nations, Statistical Yearbook, various issues); and Economic Report of the President, 1991 (Washington, D.C.: United States Government Printing Office, 1991).

permitting it to move toward a higher level of independent research.[53]

These projects began with government contracts let to private companies, which, in turn, worked with universities and public laboratories to develop a specified area. One of the first programs spawned by the Large Scale Project was the Super High Performance Electronic Computer Initiative. Between 1966 and 1981 this project received funds amounting to $45 million (in real 1982 value). Other important projects included the Pattern Information Processing System, the Direct Steel-Making Process Using High-Temperature Reducing Gas, the Comprehensive Automobile Control Technology, and the Electric Car.[54]

The institution of the Large Scale Project occurred in the wake of two important initiatives by the Prime Minister's Council for Science and Technology. In the beginning of the 1960s the Council urged the construction of a 'science city,' a place where Japan could coordinate its efforts in new and emerging science areas and ultimately nurture their growth so that it could pick its own scientific fruit. In 1963 the government selected Tsukuba as the location for this science city. Three years later construction work was under way there. By 1970 the government passed the Law for Construction of Tsukuba City for Research and Education, a measure that paved the way for the city to serve as a home for national research institutions. Two years later the first such institution, the Research Institute in Organic Materials, relocated to Tsukuba.[55]

The second major initiative of the Council for Science and Technology preceding the implementation of the Large Scale Project was that in 1965 it requested that the nation 'seek relief through indigenous research and development from over dependence on foreign technology.'[56] The Japanese not only fully understood the importance of technology for economic growth, but they clearly wanted to link science to 'in house' technology to foster development at an even faster pace. Dissatisfied with technology dependence, Japanese policy makers therefore set out in the 1960s on what was a demanding course to develop independent S&T to propel industry and the economy.

An important semi-government institution that appeared in 1967 under the direction of Monbusho (the Ministry of Education, Science and Culture) was the Japan Society for the Promotion of Science (JSPS). JSPS was an early institutional effort intended specifically to encourage science development. One of JSPS's important purposes has been to train Japan's scientific researchers. This has included the formation of cooperative committees made up of academic and industrial researchers who, through discussion, have identified next-generation developments.[57]

Sustaining the economic growth of the 1960s was very important to Japanese policy makers. Real gross national product (GNP) surged in Japan during the 1960s. Between 1961 and 1965 real GNP grew at an average annual rate of 12 percent in Japan, more than double the increase for the U.S. economy. The second half of the decade was little different; between 1966 and 1970 real GNP increased at an average annual rate of 11 percent, while in the United States the growth rate was 3 percent.[58]

Despite the strengthened economy and the efforts to develop an indigenous S&T base, technology dependence was still evident. Japan also continued to rely on the United States for military protection. Thus the need for American technology and military dependence continued to serve as major economic and political ties in the relations between the United States and Japan.

In order to fulfill its objective of strong economic growth, Japan had no alternative but to remain largely dependent on U.S. technology. In 1960 Japan's balance of payments deficit in technology know-how came to $85.5 million; by 1965 it had risen to $152.4 million.[59] Much of this deficit in technology know-how was with the United States.

Regarding the political tie, the 1960 Mutual Security Treaty reasserted America's desire to defend Japan as well as the latter's willingness to be protected. Unlike in 1951, Japanese policy makers freely signed the 1960 Mutual Security Treaty, though popular resistance was very evident. Indicating this latitude was that Japan was able to include in the treaty specific concerns important to its national interests, such as arranging for the subsequent return of Okinawa.[60] The decision to support the treaty reflected the dominant concern that the Soviet Union was a threat to Japan's national security.

Technological dependence was of course tied to Japan's objective of continued economic growth. While Japan's lofty ambition was to curtail foreign technology dependence, and this it felt could be done by developing competence in science, more realistically, for the time being, Japan had to rely on what it could do best; this was the capability to improve existing technologies. Though Japanese policy makers had made plans to launch the nation into a period where it could capitalize on the integration of indigenous science with technology, this is fundamentally different from experiencing it. For this, they had to wait.

In the meantime, the Japanese had to be content with developing a large amount of technology from the United States. Japan did, however, have one important thing going for it at the time: manufacturing exports to the United States were growing, and by the mid 1960s this included some technology-instensive goods. This gave Japan an increasingly favorable trade balance.[61]

When the United States entered the Vietnam War, Japanese industries gained additional momentum. Not only did they help serve the needs of the U.S. military, but Japanese manufacturers shifted their attention to Asian markets that American producers could not supply. Moreover, and not coincidentally, in 1965 Japan's global trade balance was in the black, something it had not experienced in more than twenty years.[62]

Also noteworthy during the 1960s was the growth and the decisive shift in Japan's foreign investments. While most of Japan's modest foreign investments in the 1950s were in the United States, during the 1960s the majority went to the Asian nations, with America dropping to third place. Asia therefore was not the only place where Japanese investors turned their sights in the 1960s. During the 1950s the Japanese put only a meager 1 percent of their investment capital in Europe. This radically changed in the 1960s. At this time over 19 percent of Japan's foreign investments were placed in European nations, only slightly less than what went to Asia.[63]

Thus Japan grew in many ways in the 1960s. Undergirding this growth was a determination to invest both in foreign and domestic markets. At the same time, the Japanese also recognized the importance of building up their export capabilities. Most significant for future economic growth and industry development was that while Japan imported more technology, it simultaneously was working to expand its S&T base. Moreover, new scientific interests appeared in Japan during the 1960s in space and nuclear research.[64]

Increasing confidence -- but continued dependence

Based on experience acquired in the past, the 1970s were years of continued development for Japan. This was a decade when economic growth remained strong, despite the fact that it followed on the heels of the phenomenal advances made in the 1960s (e.g., the *Izanagi* boom that endured for nearly five years and finally ended in the summer of 1970). However, the 1970s were also years when what was later to take the name 'Japan bashing' established roots. Considering that Japanese headlines in 1969 proclaimed the existence of a 'U.S.-Japan Textile War,'[65] the 1970s became a decade when *kembei*, the equivalent (bashing) diatribe from the Japanese standpoint, began to take shape as well. It was a decade when the Japanese steel, automobile, and machine tool[66] industries encountered trade disputes relating to the alleged unfair penetration of the U.S. market. Perhaps most important is that this was the decade when the Japanese fully realized that the collective forces of industry, government, universities, and labor could pull the nation toward an unprecedented level of development, and that this more than ever depended on both the continued advancement and integration of S&T.

The emerging bilateral crisis put Japan in a difficult and unprecedented position. On the one hand, policy makers in Japan understood that industrial advancement would cross circuits with the United States and that forgoing the opportunity to exploit the talents of the people in perhaps the world's most educated nation would be unreasonable and would directly contravene its national interests by undermining economic security. On the other hand, led by the generally U.S.-leaning Liberal Democratic Party, Japan's policy makers could not forthrightly abandon the nation's need for military security from the United States. For sure, Japan was both incapable and unwilling to provide comprehensive military protection to itself, a safeguard only the United States could afford to provide to it. Nor could Japan ignore the reality that it had developed a technological dependence on the United States and that this was still important to its industrial and economic growth. In spite of the many technological and industrial advancements made by the Japanese during the previous two decades, they were completely aware that they still had not achieved technological parity with the United States.

This contradictory position was easily reconciled by Japan. For the most part, Japan passively accepted its position in the U.S.-dominated coalition but gained increased confidence from its industrial and technological accomplishments. In the 1970s Japan was managing and applying advanced technology for industrial and general economic growth, as well as turning more of the nation's efforts toward science. Indeed, by the end of the decade Japan had established the conditions which would eventually permit it to thread science to the technological needle.

In the meantime, Japan relied on what had become routine: it continued to import foreign technology as R&D spending increased in key sectors. Real (1985 yen) R&D expenditures increased by almost 60 percent during the 1970s; those for public corporations, by nearly 200 percent, for universities, by about 50 percent, while company funds grew by nearly 50 percent.[67]

Figures 3.1 and 3.2 show Japan's technology trade in two different periods. During the 1950s and 1960s Japan's negative balance of payments in

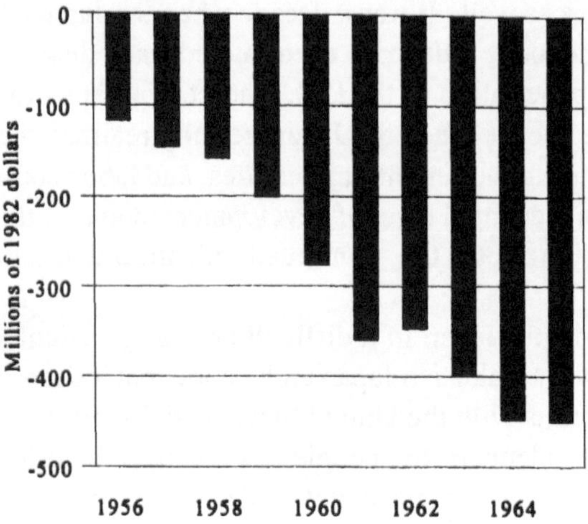

Figure 3.1 Japan's balance of payments, technology trade, 1956-1965

Sources: United Nations (U.N.) Educational, Scientific and Cultural Organization (1971); and U.N. Statistical Yearbook, various issues.

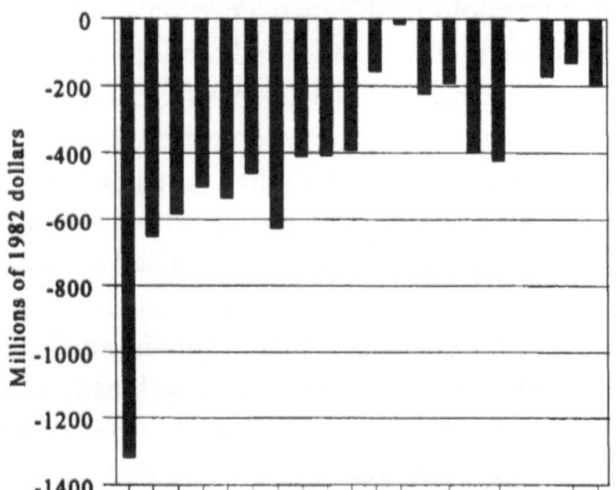

Figure 3.2 Japan's balance of payments, technology trade, 1973-1992

Sources: Organization for Economic Co-operation and Development (OECD), Basic Science and Technology Statistics (1991); and Science of Technology Agency (STA), White Paper on Science and Technology (Tokyo: The Japan Information Center of Science and Technology), various issues.

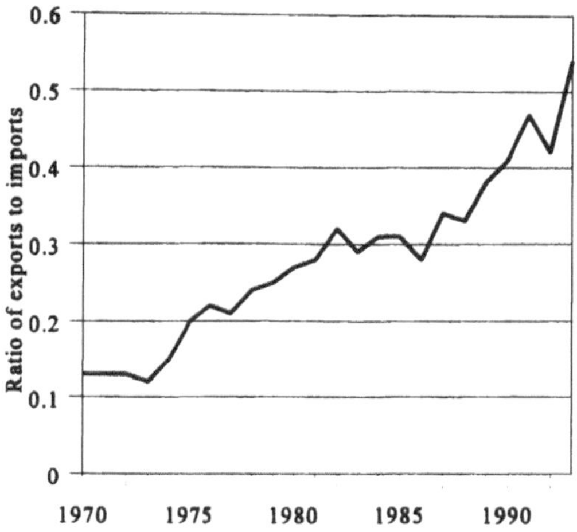

Figure 3.3 Japan's technology trade, 1970-1993

Source, STA, White Paper on Science and Technology, 1994.

technology trade or know-how (the difference between payments and receipts based on fees and royalties from patents, licensing transfers, and other types of technological capability) generally increased steadily. Shown in Figure 3.2 is the leveling off in the balance of payments deficit after 1973.

Perhaps Figure 3.3 better demonstrates the point to be made here. Japan's receipts relative to payments (i.e., the technology trade ratio of exports to imports) began to move closer together after 1973 so that even though it was still a large net importer of technology know-how to some extent it had improved its negative balance of payments. Japan did this by tripling its exports of technology know-how from 1973 to 1979. During the second half of the 1970s Japan's ratio of exports to imports in technology trade improved with many nations, including the United Kingdom, Germany, and the United States.[68]

Nonetheless, of the major industrial nations Japan by far maintained the largest negative balance of payments in technology know-how with the United States during the 1970s. In 1975 Japan's negative trade balance in technology know-how with the United States was more than 7 times larger than Canada's, more than 6 times bigger than France's, 4 times larger than Germany's, and 70 times greater than that registered for Great Britain. Indeed, between 1972 and 1979 Japan accounted for a third to more than two fifths of America's positive balance of technology trade from unaffiliated (nonAmerican) foreign sources.[69]

Moreover, Japan's purchases of U.S. technology in the second half of the 1970s were significantly higher than its agreements with other industrial nations. Between 1975 and 1979 the Japanese contracted 2,323 new technology agreements with the United States, more than double the combined amount it made with other major industrial nations. U.S. purchases of Japanese technology agreements, on the other hand, were not nearly as numerous. During the same five year period, U.S. users committed themselves to only 558 new agreements with the Japanese.[70]

For sure, Japan's use of U.S. technology was still evident in the 1970s. But by this time Japan was very skillfully adapting imported technology to its industrial needs. Exporters of American technology, on the other hand, continued to be content with reaping the capital benefits that went along with the transfer of technical know-how to Japan. U.S. corporations continued to shun Japanese technologies (as they did those of other nations), at least partly because of the widespread belief that they could learn little from foreign sources.

Japan's increased involvement in S&T -- indicated by rapid growth in R&D expenditures and the continuing consumption of U.S. technology -- during the 1970s reflects the concerns of a nation determined to become a leading economic force in the global system. Sustained attention to this objective, however, did have adverse environmental effects. Specifically, Japan's rapid economic growth during the 1960s had created some environmental problems. In the 1970s the government began to address these issues.

One of the most critical industrial developments taking place in Japan during the 1970s was the increased attention given to semiconductors. Japanese semiconductor production had only begun in the 1960s, and the sections of electronics companies doing this work were still losing money in 1972.[71] Encouraged by the generous financial and institutional support from the Ministry of International Trade and Industry, semiconductor R&D increased sharply during the 1970s. Beginning from virtually nothing in 1964, Japan's semiconductor R&D spending grew from about 9 billion yen ($37.3 million) in 1970 to 30 billion yen ($106.1 million) by 1977.[72] Accompanied by increasing capital equipment investments, the rapid growth of R&D expenditures significantly strengthened the Japanese semiconductor industry. By the end of the 1970s political salvos had already been exchanged in the U.S.-Japan 'chip war,' a technology-based conflict that for years noticeably strained the bilateral relationship.

Coaxing Japan to pay for its 'free ride'

During the 1970s important military developments took place that eventually affected Japanese S&T development. Although military R&D in Japan actually declined by 16 percent in real value (1985 yen) between 1970 and 1975, this soon changed. Top military officials from the United States and Japan began to meet on a regular basis beginning in 1975. In November 1978 both nations adopted the Guidelines for U.S.-Japan Defense Cooperation, an agreement which strengthened Japanese security as well as neighboring areas. By the end of the 1970s there was pressure from the Carter administration on Japan to increase its defense spending and share more of the military burden.[73] Meanwhile, military R&D had begun to increase in Japan. Between 1975 and 1979 real military R&D grew by 30 percent.[74] It is also important to note that since 1970 Japan's Defense Agency wanted to develop defense systems domestically rather than being dependent on the United States, although this never fully developed.[75]

As military issues were being elevated to a higher political plateau in the 1970s so were trade problems. The trade problems that emerged between Japan and the United States in the 1970s were tied to two separate posts -- Japan's evolving technological and industrial competence, and military security. However, more visible, and therefore receiving most attention, was that Japan and the United States had incompatible perspectives on trade problems, which had advanced from textiles in the 1960s, to steel, machine tools, VCRs, TVs, semiconductors, and automobiles in the 1970s.[76] Because most Japanese policy makers continued to see the nation's need for military security provided by the United States, however, they ultimately had to tolerate American complaints pertaining to the 'unfairness' of Japanese trade. At the same time they had to deal with the important matter of sharing the burden of defense spending.

As the Japanese saw it, they were efficiently managing technology, using as best they could their educated labor force in advanced production, and making some progress in science development. At the very least, all of this meant that Japan had become a technology-driven nation, with a number of industries making sizable gains in world markets, including many in the United States. But most Japanese policy makers reasoned that beyond their modest Self Defense Forces, the nation's security was largely in the hands of the U.S. military. Although Japan could have opted to end the security treaty in 1970, it chose not to do this. Instead, it involved itself in a more intense cooperative security relationship with the United States.[77]

There were manifest differences in the positions of the United States and Japan on the trade and military security issues. The Japanese public was decidedly against a rapidly growing military budget in the second half of the 1970s,[78] and therefore could only take umbrage from the U.S. criticism of their nation's need to spend more on defense. U.S. policy makers, on the other hand, largely ignored or downplayed the important technology-based industrial gains made by Japan since the 1950s. To both U.S. government officials and business leaders, that the Japanese had markedly advanced from the shambles of war did not much matter to them. Rather, they trained their attention on the unfairness of Japan's trade with the United States and the benefits it realized from getting a 'free ride' for decades from America's military security.

Although the Japanese saw America's growing economic and industrial problems resulting from gross inadequacies in government and private sector policies, these criticisms largely fell on deaf ears in the United States. By the early 1980s both policy makers and the public in the United States had made the connection between America's trade problems and the need for Japan to share more of the military burden.[79]

This connection, however, was not very meaningful and therefore smacked of political rhetoric. Had Japan increased military spending even more than it did, the trade problems with the United States would have not gone away. Japan had worked for more than two decades to strengthen its industrial base, and it relied heavily on technology to do this. That many of its industries were making appreciable market advances in the 1970s should not have been unexpected. Because the United States had no organized plan to 'redistribute' public money from military to civilian use for the purpose of revitalizing industry, a Japan taking on more of the military burden would have had no immediate ameliorating effect on bilateral trade problems. The best case scenario would have been small declines in the already big U.S. budgets deficits which may have reduced interest rates somewhat. Less upward pressure on interest rates would have been helpful, but hardly sufficient to dissolve the newly discovered, though deeply rooted, competitiveness problem.

As we have seen, Japan's peace constitution prohibited it from developing an offensive military system. Since its implementation there has always been a strong political and popular current in Japan resisting the expansion of the nation's military capabilities. Even Japan's membership in the United Nations in the mid 1950s was accompanied by a promised restriction on military activities. At the time a high level official in the Ministry of Foreign Affairs stated that 'the Japanese government will not be subject to the obligations of the United Nations Charter, which requires military cooperation or military participation.'[80] Furthermore, the existence of U.S. bases in Japan has long been a source of political controversy. For these reasons, the growth of military spending in Japan has encountered domestic obstacles. Moreover, Asian countries had long felt threatened by the prospect of a military buildup in Japan.

These constraints notwithstanding, military spending noticeably increased in Japan, though still remaining a proportionally small part of the economy. Yielding to U.S. pressure, real defense spending in Japan grew by nearly 8 percent during the latter half of the 1970s, and continued to increase, although less sharply in the 1980s.[81] Facing both domestic and foreign objections to defense spending increases, and its own fiscal crisis which restrained public spending in other areas,[82] Japan had done all that it was prepared to do to eliminate the 'free ride' label given to it by the United States.

Preparing for independence

U.S. trade problems with Japan that appeared in the 1970s were technology-based and thus had their origins in two decades of commitment to advancing industry competence. During the 1980s these problems escalated to an even higher level. Not only did America persist with the irking reminder that Japan had been receiving a 'free ride' by securing itself with U.S. military dollars, but this problem was exacerbated by the claim that Japan was also getting a 'free ride' on basic research. Japan, as the argument goes, has taken U.S. science to the market but has not sufficiently fostered much domestic science; when it has developed indigenous science, there has been restricted access to it.[83]

Although a very strong inverse statistical relationship existed through the 1980s between Japan's R&D expenditures and its negative balance of payments in technology know-how,[84] there are more details needed here to provide a complete explanation. This explanation must be prefaced by clarifying a widely held misunderstanding concerning Japan's industrial and S&T development and the connection of this to the United States. The

Japanese exploited foreign technologies by maintaining two objectives: diligence in the pursuit of improving existing technologies and consistently focusing on commercialization. This was true, for example, with the video tape recorder,[85] facsimile (fax) machines and calculators·[86] and even high speed rail systems.[87] Japan's strategies of 'learning by doing' and 'technology fusion,'[88] which help explain the nation's industrial successes, have been combined with a genuine effort to improve its educational system.

Widely recognized for its accomplishments in education, Japan's educational quality and strength has relied heavily on government initiatives. From its beginning in 1871, the Ministry of Education, Science and Culture (Monbusho) has been sensitive to the changing needs of the nation. It was, however, in the aftermath of World War II when Monbusho and the Japanese government more generally moved to modernize the nation's educational system. In doing this, government has brought increased and improved educational opportunities to the people of Japan.

A notable aside is that during the occupation period, U.S. officials wanted to dissolve Monbusho. Although this never materialized, Monbusho's powers were weakened. Having the intention of moving Japan's educational system closer to the American model, U.S. officials introduced some reforms in Japan at the time. At the end of the occupation, however, Monbusho regained its management power, and the Japanese realigned much of their educational system.[89]

In 1947 the government implemented the School Education Law and the Fundamental Law of Education. These laws created a formal and modern educational system built on the precept of equal opportunity. In 1948 upper secondary schools (for students 15 and older) were established, followed the next year by a new university system. In 1950 a temporary junior college system emerged and by 1964 it became permanent. In 1961 Japan initiated colleges of technology, with engineering as a major area of study. At the university level, the proportion of students enrolled in institutions of higher education relative to the college age population (18 to 21 years of age) grew from about 6 percent in 1950 to over 30 percent by the end of the 1980s. Recent educational reforms by Monbusho have been designed to increase the amount of education for the Japanese people and to help them adapt to the new conditions brought about by changing international demands.[90] Significant too is that Monbusho supports the largest and longest running basic research program in Japan -- the Grant-in-Aid for Scientific Research Program that began in 1965.[91]

Important to our analysis is the fact that Japan's high quality educational system has been shown to be associated with its relatively strong post-war economic growth. Japan's increasing educational attainment levels, when combined with its solid and unmitigated commitment to R&D, explain more than a quarter of the phenomenal economic growth it experienced between the 1951 and the early 1970s.[92] The overall strength of Japan's educational system, moreover, has helped prepare its work force to meet the requirements of advanced manufacturing.[93] At the higher educational levels, colleges and universities produced a continuing stream of 'corporate *samurai*,' competent industrial workers, which most definitely included engineers, scientists, and technical personnel needed to propel Japan's industrial economy.[94]

The Japanese government has also nurtured worker training programs; in this, it has been supported by industry. Unlike in the United States where most job training is for upper-level employees, worker training in Japan occurs at all occupational levels. In addition to having strong linkages between school and the workplace, firms of all sizes in Japan are heavily involved in the training of employees, and it is a continuing process.[95] Evidence from the Japanese automobile industry, a sector that has been a perennial trouble spot in bilateral trade relations, shows that Japanese production workers, supervisors, and manufacturing engineers, both experienced and inexperienced alike, received considerably more training than their counterparts in the United States.[96]

Thus a relatively strong educational system that has produced very good results beginning in the early school years has favorably complemented government- and industry-supported worker training programs and Japan's clear commitment to expanding its R&D base. These convergent efforts go a long way in explaining how Japan's well-educated and well-trained labor force has been able to advance existing technologies. They also help explain Japan's improving production efficiency and industrial competitiveness.

There has also been an effort to integrate university and industry research in Japan. In 1983 Monbusho introduced a policy statement endorsing a 'joint research system' between national universities and the private sector. Prior to this time company employees were not permitted to work in university laboratories as nonacademic researchers supported by corporate funds. Since Monbusho introduced this new policy an increasing number of university-industry projects have appeared in Japan. In 1983 there were 56 such projects with expenditures totaling almost $3 million; by 1990 there were 869 projects with spending exceeding $26 million. In 1990 these projects involved work in seven advanced-research areas: materials, machinery and equipment, computer software, civil engineering, biotechnology, electronics, and energy. Monbusho's JSPS also has a large number of cooperative academic-industrial committees involved in advanced research in areas such as superconductivity electronics, magnetic recording, and integrated transformation of electroluminescence, a breakthrough that Monbusho claims will completely transform information display technologies.[97]

In recent years Japan's unremitting dedication to advanced science and its attempt to integrate this with technology are noteworthy; at the same time, these efforts have further troubled its relationship with the United States. Although the documented erosion of American technological leadership is now widely accepted by policy makers and academic analysts,[98] many observers still see U.S. science as far ahead of the competition. This is a position which needs to be reconsidered based on developments that having been taking place in Japan.

Earlier we made clear that by the mid 1960s the Japanese had recognized the importance of promoting indigenous science. Subsequent to the recognition of this need the Japanese moved determinedly to establish a strong science base. From the information indicated in several important measurements, it is evident that Japan has broadened its science capabilities.

We have elsewhere shown the phenomenal growth of Japanese patents,[99] including those granted *within* the United States and which are steadily approaching those issued to American inventors.[100] Significantly, by 1987 Japanese companies were among five out of the top six firms receiving patents in the United States.[101] Apart from patents, the improvement in Japanese science shows up in literature analyses. There is an almost steady climb in Japan's aggregate share of publications in world scientific literature between 1973 and 1986 -- a measure that consists of eight scientific fields (clinical medicine, biomedical research, biology, chemistry, physics, earth and sciences, engineering and technology, and mathematics). Compared to other advanced nations, Japan moved from having almost the lowest share of scientific publications in 1973 to having one of the highest by 1986.[102]

In four important basic research areas the Japanese contributed more scientific papers than did researchers from the United States in 1988. In zinc selenide (possibly to be used to increase storage space for optical data), perpendicular magnetic recording (a Japanese discovery that creates the possibility for greater recording space), hot ductility of low-alloy steels (an advancement that will reduce a common cracking problem in low-alloy steels), and high performance micro-enzyme sensors (devices that will help check food quality as well as improve medical testing), Japanese university and industry researchers have established a scientific advantage over their U.S. counterparts in what are very promising commercial fields.[103]

A related development in scientific publications by Japanese authors is that through the 1980s their work has increasingly been receiving more attention. Most notably in chemistry and physics and to a lesser extent biology, Japanese scientists have experienced increased success in leading journals. Notes an editor of a major scientific journal (the *Journal of Organic Chemistry*) when comparing Japan to other nations, 'Japan has caught up in organic chemistry sooner than in other areas of chemistry.'[104]

In the fields of pharmacology, molecular biology, and biochemistry Japanese scientists produced a series of 'hot' papers between 1988 and 1990 that became visible by the extraordinarily high number of citations.[105] Moreover, with regard to the use value of scientific work, in one important area, engineering and technology, the citation ratio for Japan by the mid 1980s was virtually identical to that of the United States.[106] In organic superconductors, a field that promises at least as much as its competitor area of ceramic superconductors for advanced energy management, university scientists in Japan have made a very significant breakthrough -- producing the highest temperature using organic superconductors -- and they have also published a much larger number of papers than their U.S. counterparts.[107]

Further indicating the growing strength of Japan's science and advanced technology is the attractiveness of its universities to students. Although a very large number of students from Japan go to the United States to study, an examination of their academic interests reveals that the physical sciences (which includes chemistry, physics, and metallurgy) and engineering is not why they traverse the Pacific Ocean. Of the 18,050 Japanese students enrolled in U.S. institutions of higher education during the period 1987-1988, less than 550 of them were doing work in the physical sciences. Similarly, during this period, only 760 Japanese students were studying engineering in the United States. Moreover, only a small proportion (about 4 percent) were doing mathematics and computer science work in the United States during this period. The majority of Japanese students involved in academic work in the United States were enrolled in business management, the social sciences, and 'all other areas.'[108] Japanese students have come to appreciate that a good science, technical, and engineering education can be obtained in Japan.

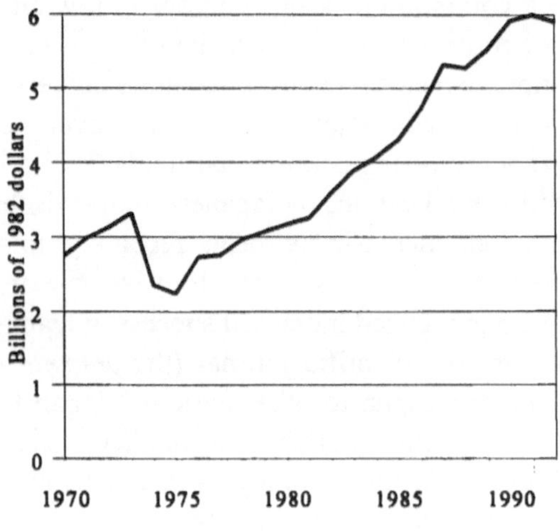

Figure 3.4 Basic research in Japan, 1970-1992

Sources: OECD, Basic Science and Technology Statistics; STA, White Paper on S&T, various issues; and OECD purchasing parity (PPP) rates.

Japan's universities have also been attracting a growing number of students from other Asian nations. In 1992 there were almost 50,000 foreign students in Japan, and most came from other Asian nations. The majority of these students were not being supported financially by the Japanese government. Nearly 30 percent of the foreign students in Japan were studying in the sciences, engineering, or medicine.[109]

Still another important indicator of Japan's strength in science is the growth of basic research. Measured in real (1982) dollars, basic research in Japan grew from $2 billion in 1975 to $5.9 billion in 1990. (See Figure 3.4.) Basic research in Japan came to 14.3 percent of total R&D spending in 1993, the highest it had been since 1980.[110] Claims that Japan has not undertaken enough basic research in the past two decades have been misleading. It is evident from Figure 3.5 that since the mid 1970s basic research to gross national product for Japan has been only slightly below that for the United States. Figure 3.6 shows that the ratio of basic research spending in Japan relative to that in the United States grew somewhat between 1975 and 1990. Thus the data indicate that Japan's basic research efforts have increased over the years; and that given the size of its economy, it has not been terribly laggard in basic research spending relative to the

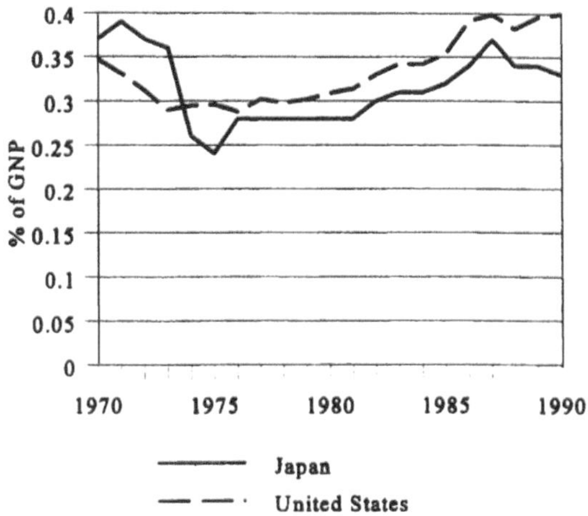

Figure 3.5 Basic research in Japan and the United States, 1970-1990

Sources: STA, White Paper on S&T, various issues; National Science Foundation (NSF) data; and National Science Board (NSB).

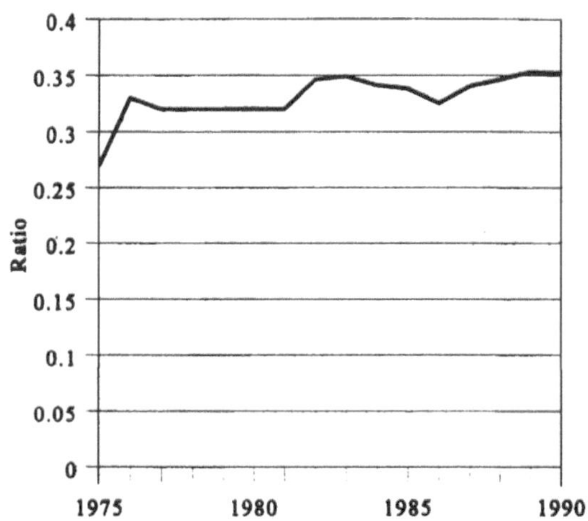

Figure 3.6 Ratio of basic research, Japan to the United States, 1975-1990

Sources: NSF; NSB; STA, White Paper on S&T, various issues; OECD's PPP rates; and Economic Report of the President (1991).

United States.

This data also indicates that the 'free ride' argument regarding Japan's dependence on U.S. science has not been properly conceptualized by critics who clearly project the impression of a paucity of Japanese basic research. To the extent that Japan has received a 'free ride' from America's basic research, it has not been so much because of the quantity of this work. This is reflected in the amount of basic research Japan has performed it relation both to the size of its economy and to the size of the U.S. economy.

The issue then is not the quantity but the *quality* of Japan's basic research. Relative to the United States, Japan has been a 'late bloomer' concerning the performance of quality basic research. This is understandable, however, since the society-wide devastation realized during World War II essentially stunted Japan's indigenous technology development and surely delayed its emphasis on the advancement of science. In the past while Japan was undertaking a reasonable quantity of basic research, it lacked the experience to perform as much quality work as that of the United States. This has begun to change. In recent years Japan has stressed both quantity and quality improvements in basic research.

Although Japan's basic research efforts have been under way for some time, it was during the 1980s, and particularly the second half of the decade, that increased concern for it became evident.[111] The first official sign that Japan was mounting an intensified basic research effort came in March 1986 when the government sanctioned the General Guidelines for Science and Technology Policy. This policy announced that 'the government should provide more R&D funds, use them more efficiently, and improve environmental conditions favorable to R&D activities in the private sector while taking steps to increase the national total of R&D investment.'[112] Because of this policy, which among other things supported '[P]romoting highly creative science and technology centered on reinforcing basic research,' seven major areas of advanced science were earmarked for development. These included electronics and information science and technology, and materials science and technology.[113] Indicating the advanced research concern of the Guidelines is that support by the Japanese government for major basic science programs increased in real (1982) value from $389 million in 1984 to $490 million in 1989.[114]

In 1987 the government of Japan enacted the Law for the Construction of Kansai Science City, the nation's second city of this kind (recall the first was Tsukuba Science City). Already located in Kansai Science City are major research organizations such as the International Institute for

Advanced Studies, the Advanced Telecommunications Research Institute International, the Ion Engineering Center, the Research Institute of Innovative Technology for the Earth, and the Graduate University for Advanced Studies of Science and Technology.[115]

In 1981 the government's Science and Technology Agency (STA) created two important basic research programs. One of these was the Special Coordination Funds for Promoting Science and Technology Program, which replaced the previously established Special Coordination Funds for Promoting Research Program. Funding for this program increased in real (1982) value from $27 million in 1984 to about $40 million in 1991. The other initiative was the Exploratory Research for Advanced Technology Program, which also received increased funding. By 1990 STA had increased its support of several other major basic research efforts, including the Human Frontier Science Program and the Frontier Research Program.

MITI's Next-Generation Fundamental Industrial Technology Research and Development Program was also established in 1981. This program had a real operating budget of over $25 million in 1984; by 1989 its expenditures had increased in real value to $27 million. Known also as the Research and Development Program on Basic Technologies for Future Industries, this MITI initiative is a cooperative effort involving government, universities and industry. Research work included in this program is expected to be connected to promising innovations for industrial application. The five concentration areas of this program are superconductivity, biotechnology, speciality materials, software, and new chip designs.[116]

Still another new initiative begun in the 1980s was the International Core System for Basic Research, or the *Shosai* Program. As a result of a report put out by the prime minister's Council for Science and Technology (CST)[117] in the Fall of 1987, STA announced the formation of the *Shosai* Program in the Spring of the following year. While one main purpose of the program is to stimulate basic research in government laboratories, another is that it requires that each project have international participation. By 1991 there were twenty-four projects in the International Core System for Basic Research Program, with a combined budget of almost $9 million dollars.[118]

Over the past few years Japan has given more attention to the miniaturization of electronics technology, or nanoscopic technologies. The Institute of Physical and Chemical Research (RIKEN) has recently begun work in this area as part of the overall efforts of the Frontier Research Program; moreover, MITI's contribution to the Large Scale National R&D Program in 1992 supports a project in this area. Important basic research in microelectronics technology is also being done by the Research and Development Corporation of Japan's (JRDC) Exploratory Research for Advanced Technology (ERATO) program, which includes the Atomcraft Project that is expected to make significant advances in semiconductor technology.[119]

It is unequivocal that the Japanese government is continuing to demonstrate its commitment to strengthening the nation's basic research. In 1992 the prime minister's CST endorsed a priority measure which included increasing basic research in Japan.[120] In particular, Monbusho and STA have worked to strengthen university research and to promote the growth of basic research.[121] The government of Japan has also recognized that it needs to work with local policy makers to foster the regional development of basic research.[122]

This recognized need to focus determinedly on basic research in Japan has indeed been one of the most glaring issues in the Japanese government's current efforts to influence the 'post-bubble economy.' As a result of the recent improvements in the 1986 General Guidelines for Science and Technology Policy, which called for the creation of Centers of Excellence to promote basic research, the STA in 1993 began providing funds to support the Centers.[123] STA has also actively supported the Promotion of Basic Research for the Exploration of New Seeds for Important New Technologies. One such effort sponsored by STA has been through its JRDC, which planned to begin funding approximately fifty 'strategic basic research projects' in the fall of 1995; other such efforts are for JRDC to strengthen the basic research work of its ERATO program and for RIKEN to do the same for its Frontier Research Projects.[124]

The serious concern of the Japanese government to advance the nation's basic research capabilities is reflected in the CST 1994 report -- Basic Plans for Promoting R&D on Advanced Fundamental Science and Technology -- to the prime minister's office. In this report the CST listed over 180 'advanced fundamental technologies' that require innovative development within a decade to push forward the existing pool of scientific and technological knowledge.[125] In 1996 STA along with five ministries, including MITI and Monbusho, implemented the New Systems for Promoting Basic Research. To support this initiative, the Japanese government provided $300 million in 1996. The biggest research program in this initiative is the Core Research for Evolutionary Science and Technology under the aegis of JRDC. Recently $50 million was made available to support this program.[126] Monbusho has also recently begun providing substantial funding for the advancement of basic research to the Japan Society for the Promotion of Science (JSPS), a quasi-government body under its direction. In 1996 Monbusho provided JSPS with $100 million to begin the Research for the Future program.[127]

Although basic research spending by government has been important to the development of science initiatives in Japan, the majority of this worked is performed at universities. Yet most striking from Japan's recent increased involvement in basic research is the rapid growth of this work performed by the private sector. Compared to the performance of basic research by U.S. corporations, work done by Japanese firms increased very rapidly during the 1980s. As we will see in Chapter 5, the quickened growth in basic research performed by Japanese companies in recent years has been a source of contention between the United States and Japan. Suffice it to say here that the desire by the United States to access this research has been a growing trouble spot in the relations between these two nations.

Turning the corner

Japan's emergence as a leading economic power, along with its major research advancements which most definitely includes significant forward strides in basic research, have helped give rise to its emphasis on the globalization of S&T. Here Japan's objectives are clear: to demonstrate the high level of very advanced S&T now being performed there, and to help deflect criticism that it does not perform enough basic research. These objectives have taken shape in the new globalization effort which broadly promotes the idea that it is incumbent for Japan to share its research results with other nations, since it has been able to demonstrate its capability to perform advanced technological and scientific work.

As we have seen above, a main part of the Shosai Program focuses on Japan developing international ties in S&T. Another major initiative demonstrating Japan's desire to globalize its S&T is MITI's plan of 'techno-globalism.'[128] MITI activated the techno-globalism plan in 1991 when it launched its International Cooperative Research Program for Intelligent Manufacturing Systems.[129] Because techno-globalism is a plan to help diffuse S&T throughout the world, it reinforces Japan's newly acquired status as a world superpower.

Dovetailing with MITI's plans of techno-globalism are other important demonstrations of Japan's recent concerns to internationalize S&T. Jointly supported by MITI and STA, the Human Frontier Science Program is an international initiative that focuses on very advanced work in the life sciences. This program both subsidizes collaborative international research and supports scientists desiring to work in a foreign nation. First suggested by the Japanese government in June 1987 at the Venice Economic Summit, by March 1991 this program was supporting the work of 232 scientists from twenty two different nations.[130] Other examples of international programs begun by the Japanese government in recent years are the STA's Fellowship Program, a fellowship program supported by the JSPS, and the Agency of Industrial Science and Technology's Foreign Researcher Invitation Program.[131]

Thus calls for the global sharing of research as well as more basic science indicate Japan's determination to assume a superpower position. At the same time, these activities help nudge Japan into a leadership position in global politics. This push by the Japanese government to promote more basic research, including increased public spending for this purpose, becomes noticeably more cautious, however, when the topic turns to accessing the work of private companies.[132]

Further indicating Japan's determination to mold for itself a superpower position based on advanced S&T is the recent changes in its national policy. Revisions in the original (1986) General Guidelines for Science and Technology Policy mandated in April 1992 announce the government's desire 'to increase the number of engineers, researchers and research assistants... to double its own R&D investment as early as possible... and to promote basic science [which] expands the intellectual frontier, forms the foundation for creating new views of nature, gives hope to the people, and suggests new ideas and guidelines for the science and technology of future generations.'[133]

The recession that Japan experienced in the early 1990s hit R&D expenditures hard. Between 1991 and 1992 Japan's total R&D spending grew by only 1 percent, the smallest increase in almost four decades. Also without precedent since the early 1950s, corporate spending declined by nearly 2 percent in 1992. Partly reflecting the support of government to S&T advancement in Japan, research spending by both universities and research institutes grew between 1991 and 1992, the former by 7 percent and the latter by over 9 percent.[134] Still the bursting of the *baboru keizai* (bubble economy) did have an overall damaging effect on R&D; without precedent, total R&D spending in Japan declined in both 1993 and 1994.[135] More recently, there has been serious consideration given to the reinvigoration of Japanese R&D spending.

While the United States has consistently sidestepped environmental problems and therefore has certainly not occupied a world leadership position in this area, Japan has been expanding its involvement. In 1989 the Japanese government formed the Council of Ministries for Global Environmental Conservation. The next year this council created the 'Action Program to Arrest Global Warming' which is scheduled to continue into the next century. The objectives of this program are to put Japan's economic and S&T resources to work on the problem of global warming and to involve the international community. Japan has also demonstrated increased interest in other environmental matters such as acid rain, ozone protection, and deforestation. Indicating its growing concern with world problems in this area, the government's Environment Agency formed a Global Environment Department in 1990; the Environment Agency has also created the Center for Global Environmental Research.[136]

The government of Japan is quite supportive of international organizations concerned with the environment. Two examples of this are Japan's involvement in the United Nations Environment Program and the Organization for Economic Co-operation and Development's Committee on the Environment. In 1991 Japan gave over $16 million to the United Nations Environment Program. This program will implement an International Environmental Technology Center to help underdeveloped countries deal with environmental problems; the center will be maintained by Japanese funds.[137]

MITI alone was appropriated $893 million for the environment in 1991. One of MITI's concerns in this area is to develop new industrial technologies to solve global environmental problems. By drawing on the assets of universities, industry, and government, MITI plans to form the Research Institute of Innovative Technology for the Earth.[138]

The government of Japan is unquestionably taking global environmental problems very seriously. Spending for this purpose has been increasing during the 1990s. Figures from Japan's Environment Agency, show real dollar growth for global environmental work between 1991 and 1992.[139] More recent figures show continued growth in the Agency's global environmental projects, including increasing spending for S&T support.[140]

Japan in retrospect and in prospect

We have shown in this chapter that Japan's S&T base evolved from almost complete dependence on the United States to what could now be characterized as relative independence. It is completely wrong to assume -- as most Western analysts have -- that Japanese technology and science development was simply the result of foreign transfers, that is to say, from the United States. Japan's technology development, including that in myriad high technology areas involving both product and process technologies, incontrovertibly required substantial indigenous effort. This effort, for example, evident in the Japanese production of steel, reflected an unwavering commitment to incremental engineering developments and to generalized manufacturing improvement.[141]

While Japan still imports more technology know-how from the United States than it exports there, there definitely has been considerable improvement over the years. Japan's ratio of exports to imports in technology know-how (or trade) with the United States increased from 0.19 in 1981, to 0.25 in 1985, to 0.38 by 1992. Moreover, figures from the Management and Coordination Agency showed that the ratio of Japan's technology trade with all nations had been at or close to 1 in the early 1990s.[142] In 1993 a watershed in Japan's technology trade balance occurred. The Management and Coordination Agency reported in its 1994 R&D survey that the ratio of payments to receipts in technology trade was 0.91, making 1993 the first time that Japan's income exceeded its payments in this area.[143]

In the early days Japan imported technology, methodically developed it further, and most definitely put it to productive use. In doing this government became the midwife of technology and industrial development in Japan. In time Japan's interest in developing an indigenous S&T base became evident. This occurred concurrently with the trade disputes between itself and the United States; however, the development of Japan's domestic S&T infrastructure certainly was then not an integral part of the controversy. What did become important at that time was the concern in the United States with 'burden sharing,' and the linkage of this to the emerging trade disputes.

Steady improvements in Japan's technological capabilities were met with increased pressure by the United States to acquire access to it. During the 1980s it became apparent the that bilateral technological table was beginning to turn, as the United States increased pressure on Japan for access to its research and technology. As we shall see in Chapter 5, access to S&T is not problem free.

As Japan's science capability has markedly improved, as has the promise of integrating this effectively with its already very strong technology base and its unparalleled commercialization experience, it has consequently become increasingly concerned with global issues. Japan's newly acquired superpower status, however, has put it in a position that at times contrasts sharply with the United States. Because S&T will become more and more important as we approach and then enter the next century, it will have an increasingly significant part in world politics. To the extent that Japan's S&T base continues to evolve differently from that in the United States, the prospect of diverging interests rather than concordant ones will predominate relations between the world's two superpower.

Notes to chapter 3

1. See Department of State, *Occupation of Japan: Policy and Progress*, (Washington, D.C.: United States Government Printing Office [USGPO], 1946), pp. 41-42; and Alun Anderson, *Science and Technology in Japan* (Essex, U.K.: Longman, 1984), p. 109.

2. Japanese National Commission for UNESCO (United Nations Educational, Scientific and Cultural Organization, *Technological Development in Japan* (Paris: UNESCO, 1971), p. 40.

3. See Tetsuro Nakaoka, 'A Giant Home Electronics Company in a Declining Textile Region: Postwar Strategy of Matsushita Denki (Matsushita Electric Industrial Company)'; Takeshi Abe, '*Obitani Shoten*: The Strategy and Structure of a Cotton Weaving Firm in Sen' nan District of Osaka Prefecture'; Takeshi Abe and Jun Sasaki, 'The Change in Big Business in Osaka Prefecture'; Kingo Tamai, 'The Post-War I Structure of Industrial Relations in Osaka: The Case of Textile Industry'; and Kingo Tamai, 'A Historical Review of Industrial Relations in Osaka: The Case of Textile Industry.' Papers presented at Business Strategies in Restructuring Textile Regions in Great Britain, Japan, and the United States; a conference at the Hagley Museum and Library, Wilmington, Delaware, September 18, 1993.

4. Tuge Hideomi, ed. *Historical Development of Science and Technology in Japan* (Tokyo: Japan Cultural Society, 1968), pp. 124, 128.

5. Ibid., 129; and UNESCO, *Technological Development in Japan*, p. 142.

6. UNESCO, *Technological Development in Japan*, pp. 41-42.

7. See Nakaoka, 'A Giant Home Electronics Company in a Declining Textile Region: Postwar Strategy of Matsushita Denki.'

8. See Department of State, *Occupation of Japan*, pp. 41-42, 165; and Tuge, *Historical Development*, pp. 143-144.

9. Department of State, *Occupation of Japan*, pp. 11-12, 85-88.

10. Article 9 of the new constitution specifically stipulated that Japan would no longer be a military threat to the world.

11. See Tamai, 'The Post-War I Structure of Industrial Relations in Osaka: The Case of the Textile Industry'; Tamai, 'A Historical Review of Industrial Relations in Osaka: The Case of the Textile Industry'; and Theordore Cohen, *Remaking Japan: The American Occupation as New Deal* (New York: Free Press, 1987).

12. Quoted in Department of State, *Occupation of Japan*, p. 40.

13. UNESCO, *Technological Development in Japan*, p. 42.

14. Richard Rosenbloom and Michael Cusumano, 'Technological Pioneering and Competitive Advantage: The Birth of the VCR Industry,' *California Management Review*, Summer 1987, p. 57.

15. See Tuge, *Historical Development*, pp. 142, 156, 164-165, 173.

16. See Akira Iriye, 'Japan's Defense Strategy,' in *The Annals of the American Academy of Political and Social Science*, January 1991, p. 40.

17. G. R. Hall and R.E. Johnson, 'Transfers of United States Aerospace Technology to Japan,' in *The Technology Factor in International Trade*, ed. Raymond Vernon (New York: National Bureau of Economic Research, 1970), p. 313.

18. Anderson, *Science and Technology in Japan*, p. 358.

19. See G. C. Allen, *A Short Economic History of Modern Japan* (New York: Praeger, 1963), p. 214. Also see Cohen, *Remaking Japan*, p. 439; and Robert Angel, *Explaining Economic Policy Failure: Japan in the 1969-1971 International Monetary Crisis* (New York: Columbia University Press, 1991), p. 39.

20. See Tuge, *Historical Development*, pp. 149, 175; and Yamauchi Toshihiro, 'Gunning for Japan's Peace Constitution, *Japan Quarterly*, April-June 1992, p. 159.

21. Such an argument is made in Daniel Spencer, 'An External Military Presence, Technological Transfer, and Structural Change,' *Kyklos*, vol. XVIII (1965), pp. 451-473.

22. UNESCO, *Technological Development in Japan*, p. 42; and Toshio Shishido, 'Japanese Industrial Development and Policies for Science and Technology,' *Science*, January 1983, p. 261.

23. Hall and Johnson, 'Transfers of United States Aerospace Technology to Japan,' pp. 323-338.

24. Richard Storry, *A History of Modern Japan* (New York: Penguin, 1960), pp. 251, 258.

25. See Tuge, *Historical Development*, p. 178.

26. Cohen, *Remaking Japan*.

27. Anderson, *Science and Technology in Japan*, pp. 24-26.

28. UNESCO, *Technological Development in Japan*, pp. 43-49.

29. Ministry of International Trade and Industry (MITI), *Issues and Trends in Industrial/Scientific Technology: Towards Techno-Globalism*, June 1992.

30. Shishido, 'Japanese Industrial Development,' p. 261.

31. Anthony DiFilippo, 'Why Is Japan Winning the High-Tech Battle?' *Forum for Applied Research and Public Policy*, Fall 1992, p. 18; United

States Congress, Office of Technology Assessment, *Commercializing High-Temperature Superconductivity* (Washington, D.C.: USGPO, June 1988), pp. 45-47; and Rosenbloom and Cusumano, 'Technological Pioneering,' pp. 51-56.

32. Rosenbloom and Cusumano, 'Technological Pioneering,' p. 56.

33. Shishido, 'Japanese Industrial Development,' pp. 259-261.

34. See Clyde Prestowitz, *Trading Places* (New York: Basic Books, 1988), pp. 33, 219.

35. Tuge, *Historical Development*, p. 176.

36. See Anderson, *Science and Technology in Japan*, pp. 26, 265.

37. Science and Technology Agency (STA), *White Paper on Science and Technology, 1990* (Tokyo: The Japan Information Center of Science and Technology, 1991), p. 167.

38. Tuge, *Historical Development*, pp. 172-173.

39. Anderson, *Science and Technology in Japan*, p. 26.

40. See Cecil Uyehara, 'Appraising Japanese Science and Technology,' in *Japan's Economic Challenge* (Washington, D.C.: USGPO, October 1990), p. 291 (Study Papers Submitted to the Joint Economic Committee, Congress of the United States).

41. See Tuge, *Historical Development*, p. 166 where it is reported that the number of scientists and technicians traveling from Japan to the United States increased from 65 in 1949, to 265 in 1950, to 320 in 1951, to 717 in 1952.

42. Statistics Bureau, Management and Coordination Agency, *Historical Statistics of Japan* (Tokyo: n.d.).

43. STA, *White Paper, 1990*, p. 274.

44. Ibid., p. 273.

45. Anderson, *Science and Technology in Japan*, pp. 26-27, 73.

46. OTA, *Commercializing*, p. 74.

47. OTA, *The Big Picture: HDTV & High-Resolution Systems* (Washington, D.C.: USGPO, June 1990), p. 27.

48. STA, *White Paper, 1990*, p. 274.

49. OTA, *Competing Economies: America, Europe, and the Pacific Rim* (Washington, D.C.: October 1991), p. 243.

50. STA, *White Paper, 1990*, pp. 268-269.

51. Anderson, *Science and Technology in Japan*, p. 27.

52. MITI, *Trends and Future Tasks in Industrial Technology*, September 1988.

53. STA, *White Paper, 1990*, p. 282.

54. Information received from Japanese Embassy, Washington, D.C. on the National Research and Development Program -- The Large Scale Project.

55. See Anderson, *Science and Technology in Japan*, pp. 365, 368; and also STA, *White Paper on Science and Technology, 1991* (Tokyo: The Japan Information Center of Science and Technology, 1992), p. 225. By 1992 a total of 47 operations were already working or were almost completed in Tsukuba Science City, excluding private efforts.

56. National Science Foundation (NSF), *The Science and Technology Resources of Japan: A Comparison with the United States* (NSF 88 318), p. 12.

57. Ministry of Education, Science and Culture, *Research Cooperation Between Universities and Industry: Cultivating Our Future by Fostering Creative, Progressive Research* (Tokyo: Science and International Affairs Bureau, n.d.).

58. *Economic Report of the President* (Washington, D.C.: USGPO, 1991), p. 411.

59. UNESCO, *Technology Development in Japan*, p. 43. Yen converted to dollars from exchanges rates found in United Nations, *Statistical Yearbook* (New York: United Nations, 1966 and 1970 issues).

60. See United Nations Association of the USA and the Asia Pacific Association of Japan (hereafter, U.N. Association), *The Japanese-American Alliance: A Framework for the Future* (New York: United Nations Association of the United States of America, 1983), pp. 11-12; and James Auer, 'Defense Burdensharing and the U.S.-Japanese Alliance,' in *Japan and the United States: Troubled Partners in a Changing World* (New York: Brassey's, 1991), pp. 29, 31 (Special Report).

61. See Anthony DiFilippo, *From Industry to Arms: The Political Economy of High Technology* (Westport, CT: Greenwood Press), for information and discussion on the U.S. high technology trade deficit with Japan, pp. 23-24.

62. Iriye, 'Japan's Defense Strategy,' p. 42.

63. Ryutaro Komiya and Ryuhei Wakasugi, 'Japan's Foreign Direct Investment,' p. 52, in *The Annals of The American Academy of Political and Social Science*, January 1991, *Japan's External Economic Relations: Japanese Perspectives*.

64. Anderson, *Science and Technology in Japan*, p. 28.

65. Quoted in Kenjiro Nishi, 'Developing a Thicker Skin,' *By the Way*, p. 32, January/February 1993.

66. See Anthony DiFilippo, *Military Spending and Industrial Decline: A Study of the American Machine Tool Industry* (Westport, CT: Greenwood Press, 1986), Chapter 5.

67. Management and Coordination Agency, *Historical Statistics of Japan*.

68. STA, *White Paper on Science and Technology, 1991*, pp. 147, 247.

69. Calculated from NSF, *International Science and Technology Data Update: 1991* (NSF 91-309), p. 117.

70. National Science Board (NSB), *Science and Engineering Indicators - 1989* (NSB 89-1), pp. 383-384. Between 1980 and 1988 this trend continued. During this period Japanese sales agreements with the United States totaled 2,179, and purchase agreements came to 4,965. More recent data in NSB, *Science and Engineering Indicators - 1991* (NSB 91-1) p. 414.

71. OTA, *Competing Economies*, p. 249; and OTA, *Contributions of DOE Weapons Labs and NIST to Semiconductor Technology* (Washington, D.C.: USGPO, September 1993), p .57.

72. Anderson, *Science and Technology in Japan*, pp. 28-29.

73. U.N. Association, *The Japanese-American Alliance*, p. 17; and Auer, 'Defense Burdensharing and the U.S.-Japanese Alliance,' pp. 32-33.

74. Calculated from STA, *White Paper, 1991*, pp. 236, 253.

75. Mike Mochizuki, 'U.S.-Japan Economic Relations in the 1990s,' in *Japan and the United States: Troubled Partners in a Changing World* (New York: Brassey's [U.S.], 1991 Special Report); and Richard Kaufman, 'The Japanese Defense Sector in Perspective,' in *Japan's Economic Challenge*, p. 372.

76. *Japan Economic Almanac 1991*, (Tokyo: *Nihon Keizai Shimbun*, 1991), p. 32.

77. Auer, 'Defense Burdensharing in the U.S.-Japanese Alliance,' p. 31.

78. Larry Niksch, 'Defense Policy,' in *Japan's Economic Challenge*, p. 358.

79. U.N. Association, *The Japanese American Alliance*, pp. 17-20; also see Japan Economic Institute (JEI) of America, *Yearbook of U.S.-Japan Economic Relations in 1981* (Washington, D.C.: JEI, 1981), pp. 49-50; and JEI, *Yearbook of U.S.-Japan Economic Relations in 1982*, pp. 68-69.

80. Quoted in Yamauchi, 'Gunning for Japan's Peace Constitution,' p. 161.

81. Kaufman, 'The Japanese Defense Sector in Perspective,' p. 376.

82. JEI, *Yearbook of Japan-U.S. Economic Relations in 1983*, p. 24.

83. See MITI, *Issues and Trends in Industrial/Scientific Technology*, p. 2; and NSF, *Tokyo Report Memorandum* (Tokyo: NSF April 30, 1991), p.

1.

84. For the years 1980 through 1990 the coefficient of determination, R^2, measuring the relationship between Japan's research and development spending and its negative technology trade (know-how) balance (Bank of Japan figures) is .932475.

85. See Rosenbloom and Cusumano, 'Technological Pioneering,' pp. 51-76.

86. F. M. Scherer, *International High-Technology Competition* (Cambridge, MA: Harvard University Press, 1992).

87. OTA, *Commercializing*, p. 74.

88. See Fumio Kodama, *Analyzing Japanese High Technologies: The Techno-Paradigm Shift*, (London: Pinter, 1991); and Fumio Kodama, 'Technology Fusion and the New R&D,' *Harvard Business Review*, July-August 1992, pp. 70-78.

89. Ikuo Amano, 'Japanese Schooling in Historical Perspective,' *Economic Eye: A Quarterly Digest of Views From Japan*, Spring 1993, pp. 3-8.

90. Ministry of Education, Science and Culture, *Monbusho: 1990* (Tokyo: Ministry of Education, Science and Culture, 1990).

91. STA, *White Paper, 1990*, p. 10.

92. See Hiroshi Doihara, 'Economic Lessons: Education Plays a Vital Role in Growth of Japan's Economy,' *By the Way*, November/December 1994, p. 18.

93. Chalmers Johnson, 'Economics and the Classroom: How Japan Measures Up,' *Forum for Applied Research and Public Policy*, Winter 1989, pp. 48-56.

94. Hisao Oka, 'The Industrial Sector's Expectations of Engineering Education,' *Economic Eye*, Spring 1993, pp. 12-13.

95. OTA, *Worker Training: Competing in the New International Economy* (Washington, D.C.: USGPO, September 1990).

96. Two examples in the automobile industry for newly hired employees are that production workers received 315.5 hours of training in Japan, compared to 45.7 in the United States, and manufacturing engineers received 156.7 hours of training, while their U.S. counterparts received 72.3 hours. See data in OTA, *Competing Economies*, p. 103.

97. See Ministry of Education, Science and Culture, *Research Cooperation Between Universities and Industry*; and data sheet supplied by the Japanese Embassy, Washington, D.C. Also see Kenkichiro Koizumi, 'Historical Turning Points in Japanese Joint Research Policy,' *Science and Public*

Policy, October 1993, pp. 313-314.

98. NSB, *The Competitive Strength of U.S. Industrial Science and Technology Strategic Issues* (NSB 92-138); and F. M. Scherer, *International High-Technology Competition.*

99. DiFilippo, *From Industry to Arms*, pp. 37-43.

100. See Anthony DiFilippo, 'Converting U.S. Science and Technology Resources to Civilian Use: Industrial Exigencies and Political Implementation,' in *Conversion: Opportunities for Development and Environment*, eds. A. Brunn, L. Baehr, and H.-J. Karpe (Berlin: Springer-Verlag, 1992), pp. 268-281.

101. STA, *White Paper, 1991*, p. 60.

102. NSF, *International Science and Technology Data Update: 1991*, p. 92.

103. Institute for Scientific Information, *Science Watch,* February 1990, pp. 1-2.

104. Institute for Scientific Information, *Science Watch*, May 1990, p. 7.

105. Institute for Scientific Information, *Science Watch*, April 1991, pp. 1-2.

106. NSF, *International Science and Technology Data Update: 1991*, p. 97.

107. Institute for Scientific Information, *Science Watch*, 1989, p. 7.

108. See NSF, *International Science and Technology Data Update: 1991*, pp. 65, 67; and Anthony DiFilippo, 'Working With Japan to Revitalize U.S. Industry, *IEEE Technology and Society Magazine*, Winter 1993, pp. 21-27.

109. Dennis Normile, 'Japanese Universities Become Magnets for Asian Students,' *Science*, October 1993, p. 351.

110. NSF, *Tokyo Report Memorandum*, December 6, 1994; and STA, *White Paper on Science and Technology, 1994* (Tokyo: The Japan Information Center of Science and Technology, 1995), Appendix.

111. Chihiro Watanabe and Yukio Honda, 'Japanese Industrial Science and Technology Policy in the 1990s: MITI's Role at a Turning Point,' *Japan and the World Economy* 4:1 (1992): 47-67.

112. NSF, *Tokyo Report Memorandum*, May 6, 1992, p. 1.

113. STA, *White Paper: 1991*, p. 159.

114. STA, *White Paper: 1990*, p. 10. This is an increase of 26 percent.

115. NSF, *Tokyo Report Memorandum*, March 29, 1991.

116. See STA, *White Paper, 1990*, pp. 9-11, 283-284; and STA, *White Paper, 1991*, pp. 233-234. Yen to dollar conversion done by using the Organization for Economic Co-operation and Development's purchasing

power parity index and GNP deflators.

117. Created in 1959, the CST is under the direction of the prime minister and is the government's chief S&T planning organization.

118. NSF, *Tokyo Report Memorandum*, September 13, 1989; and NSF, *Tokyo Report Memorandum*, May 24, 1991.

119. NSF, *Tokyo Report Memorandum*, January 6, 1993.

120. NSF, *Tokyo Report Memorandum*, June 6, 1994, p. 2.

121. NSF, *Tokyo Report Memorandum*, March 14, 1994, p. 2.

122. STA, *White Paper on Science and Technology, 1992* (Tokyo: Science and Technology Agency, September 1992, Summary), p. 12.

123. NSF, *Tokyo Report Memorandum*, February 3, 1995.

124. NSF, *Tokyo Report Memorandum*, October 31, 1995.

125. NSF, *Tokyo Report Memorandum*, February 27, 1995.

126. NSF, *Tokyo Report Memorandum*, (Report #96-15) July 3, 1996.

127. NSF, *Tokyo Report Memorandum*, (Report #96-16) July 3, 1996. The seriousness attached to Japan's commitment to advance science is most recently reflected in a plan by prime minister Hashimoto's Cabinet calling for a marked growth in science expenditures between 1996 and 2000. See 'Japan to Spend More on Science,' *Associated Press*, accessed on World Wide Web, July 2, 1996.

128. See MITI, *Issues and Trends in Industrial/Scientific Technology*.

129. *Japan Economic Almanac 1991*, p. 77.

130. *Japan Economic Almanac 1990* (Tokyo: *Nihon Keizai Shimbun*, 1990), p. 94; and STA, *White Paper: 1991*, p. 232.

131. See STA, *White Paper: 1991*, p. 233 for other Japanese programs supporting international exchange; also see NSF, *Tokyo Report Memorandum*, January 27, 1993.

132. See Anthony DiFilippo, 'After Paternalism: Science, Technology and the Fissuring U.S.-Japan Coalition,' paper presented at the Allied Social Science Association Meeting, Anaheim, CA, January 6, 1993; and MITI, *Issues and Trends in Industrial/Scientific Technology*, p. 40.

133. NSF, *Tokyo Report Memorandum*, May 6, 1992.

134. STA, *STA Today*, February 1994, p. 8.

135. Management and Coordination Agency, Statistics Bureau, *Report on the Survey of Research and Development (*Tokyo: Management and Coordination Agency, 1996).

136. See Government of Japan, *Action Program to Arrest Global Warming* (Decision made by the Council of Ministers for Global Environmental Conservation), October 23, 1990; Ministry of Foreign

Affairs, *Japan's Environmental Endeavors* (n.d.); and Environment Agency, *Environmental Protection Policy in Japan* (Tokyo: Environment Information Center, 1991).

137. See Environment Agency, *Environmental Protection Policy in Japan*. This is in addition to the substantial financial assistance provided by the Japanese government to underdeveloped countries.

138. NSF, *Tokyo Report Memorandum*, January 30, 1992, p. 11.

139. NSF, *Tokyo Report Memorandum*, January 15, 1993, pp. 12, 17.

140. NSF, *Tokyo Report Memorandum*, January 20, 1994.

141. See Takeshi Hayashi, *The Japanese Experience in Technology: From Transfer to Self-Reliance* (Tokyo: United Nations University Press, 1990), pp. 32-33.

142. STA, *White Paper on Science and Technology, 1992* (Tokyo: Japan Information Center of Science and Technology, 1993), pp. 260-261; and STA, *STA Today*, February 1994, p. 8.

143. NSF, *Tokyo Report Memorandum*, December 6, 1994.

4 How to holler foul: making trade policy in an unequal alliance

In an important way the 1980s marked a critical turning point in U.S.-Japan relations. Acquired expertise in many science-based technologies enabled several Japanese manufacturing industries to become considerably stronger than they had been in the past. This, in turn, provided Japan with the political muscle to begin the difficult task of attempting to eliminate the misperceived image that it could not compete fairly. The difficulty of this task was that because the Japan-U.S. alliance had not developed equally, Japan was under constant pressure to prove itself to the United States, and also to other industrialized nations. On the one hand, Japan had to demonstrate convincingly that its exports to the United States and to the rest of the world had increasingly become quality-made products. On the other hand, since it was competing head on with the United States, it had to contend with the political repercussions stemming from an unequal alliance.

By relying on a policy tool first used in the 1950s, the United States began to play up the 'unfairness' of Japan's manufacturing trade, which ultimately came to include constant charges that its domestic market was closed to many American products. Because Japan has matured, however, it has become progressively less willing to accept completely American allegations. These complications have led to increasing bilateral friction. As Japan's confidence has grown in the 1990s, bilateral relations have been further strained.

This chapter will analyze the 'unfairness' theme that has been an integral disturbing force in U.S.-Japan relations. It will argue that Japan's alleged unfair trade regime is much more of a political contrivance on the part of the United States than an established practice. The available evidence does not convincingly indicate that Japan's merchandise trade markets are closed to foreign producers.

Backdrop

Two decades ago Japan was seen as an emerging industrial nation that could not seriously challenge U.S. industrial and technological might. Now, however, it is generally accepted that Japan is an economic and science and technology (S&T) superpower, and today America's main revival. Were it not for the substantive steps taken by policy makers in Washington and Tokyo that effectively managed bilateral trade, Japanese manufacturing industries would have experienced even more success than they have already. Indeed, one could make the argument that Japan's recent recession was exacerbated by the bilaterally agreed restraints placed on its exports to the United States.

Japan's commercial achievements are the result of a sustained national effort to build a strong economy with a sound and steadily advancing S&T infrastructure. Important to this effort has been a coordinated industrial policy with support from universities and national laboratories and the resolve to include labor. We therefore should seriously question arguments that maintain that Japan's accomplishments have occurred mainly through unfair trade linked to a conspiracy ideology and excessive pay to lobbyists,[1] or because of other improper activities[2] -- at least no more so than that practiced by American industry.

That Japan has become a superpower because it has directed much of its national resources to developing what is now a sophisticated S&T base is incontrovertible. However, in the United States this is too easily dismissed as not as significant as the unfair trade practiced by the Japanese. In the United States, moreover, the risk-taking task assumed by the Japanese government has been anathematized by most observers, causing it to be easily incorporated into the unfairness issue. Although American policy makers have acknowledged that technological innovations are the main driving forces of economic change,[3] like most other observers in the United States, they have commonly downplayed the ongoing contributions made by Japan's coordinated national S&T policy to its production efficiency and trade capability.

It is often said that Japanese industrial development has followed the Schumpeterian model which does not reject government intervention and which advocates increasingly large doses of technology to improve production efficiency and national competitiveness.[4] Since the end of World War II the Japanese have developed industrial and technological policies which have continued to be directed at the commercial sector.[5] Not only has the overall effectiveness of these policies been minimized by most U.S. observers, but they are frequently seen as being unsuitably adaptable to the free market engines powering American industry.[6] Moreover, Japan's different policies have become the gist of the hard-line 'revisionist' position which stresses that it is most unlikely that Japan will adopt Western business and policy-related standards.

When Japan emerged as a major industrial and technological force in the 1980s many arguments appeared attempting to explain U.S. trade and high technology problems. This had the effect of relegating Japanese S&T and other industry policies to a relatively peripheral part of the overall explanation, which typically included some suggestion of unfairness. One of these arguments was that the high compensation paid to American workers was said to be a major reason for the waning competitiveness of U.S. industry; too much capital, it was argued, went to pay high wages and other benefits, leaving too little for new productive investments on machinery, equipment, and research and development. However, not only has total worker compensation in foreign industrial nations, including Japan, rapidly approached that of the United States, but notable increases in Japanese manufacturing investment in America in recent years further erodes the plausibility of the high wage argument.[7]

Although we have dealt effectively with this and other arguments claiming to account for America's manufacturing and high technology problems elsewhere,[8] the escalating trade problems between the United States and Japan in the 1990s has maintained the attractiveness of the unfairness issue. While the Japanese government's supporting policies have been interpreted as creating an unfair advantage for manufacturers, leading to the criticism that it has consistently donned an anti-free market suit that conspicuously clashes with its claimed capitalist anatomy, the most visible attacks of unfairness by U.S. observers have been directly linked to trade. Thus one aspect of this relates to Japanese exports to the United States, and focuses specifically on 'dumping,' that is, selling below market value, and the other concentrates on the restricted access of American business to Japanese markets.

To many observers in the United States it became apparent that the commercial threat coming from Japan was not a temporary problem that would soon go away, primarily because the Japanese simply would not practice fair trade. The continuing flood of Japanese imports to the United States which became very evident in the 1970s, had become only part of the problem by the 1980s. Soon after taking office, members of the Reagan administration began to place increasing emphasis on the lack of access to Japan's domestic markets.[9] So when the U.S. manufacturers trade deficit with Japan increased by nearly 50 percent between 1983 and 1984, exceeding $44 billion, over a third of which was made up of high technology products,[10] strident claims of closed Japanese markets combined conveniently with unwarranted government intervention and the dollar appreciation argument in an attempt to explain, however inadequately, America's trade problems.

To be sure, some suspected that America's weakened competitiveness ran much deeper than trade problems, given the string of reports and studies completed both in and out of government indicating such things as relatively poor productivity growth, a low savings rate, and insufficient investment in capital equipment and civilian research and development. However, these and other serious economic problems were left largely unattended because they required the substantial involvement of government to begin to rectify them. Broadening and intensifying the unfairness theme thus became the U.S. policy toward Japan because it required government management in bilateral discussions and the use of a 'get-tough approach,' and not direct and sustained intervention in the domestic economy.

Some of what American analysts and policy makers observed and did not like related to the 'different' and 'peculiar' ways in which Japan's private and public sectors worked together.[11] From the American perspective, the government of Japan, and particularly the Ministry of International Trade and Industry (MITI),[12] was far too willing to assist private sector development. Although U.S.-Japan bilateral trade problems became very noticeable during the 1980s, Japan's evolving industrial policy reached late adolescence in the early 1970s. At about the same time the independent advancement of high technology industries became more important to Japan's industrial policy.[13] Led by MITI and supported by other government agencies, the advancement of S&T became a national imperative. Under the guidance of MITI, the Agency of Industrial Science and Technology (AIST) increased its efforts to coordinate and advance industrial technology. Beginning in the late 1970s the Japanese government moved toward *itakuhi* financing, a policy of providing all of the funds for risky R&D projects.[14] Japanese high tech manufacturers, with government assistance, worked determinedly throughout the 1970s to improve industrial technologies, many of which were first developed in the United States.

At the end of the decade, Japanese manufacturing competence, which included many high technology products, was evident. By 1980 in some high tech areas with considerable growth potential, such as video tape recorders, U.S. manufacturers had no chance of competing successfully with Japanese producers.[15] In many other areas, all that was lacking was the opportunity to expand. Dollar appreciation relative to the yen in the early 1980s helped provide just such an opportunity, especially since the image of American-made products had become marred by poor quality and rising prices. Having already been conditioned to Japanese high tech products for some time, American consumers, despite some lingering resistance to buying foreign, saw Japanese products as a quality substitute.

Mainly for these reasons, in the second half of the 1980s America's trade deficit with Japan grew worse. Since currency realignment did not correct America's trade problem, U.S. policy makers and analysts trained more of their attention on closed Japanese markets, which was exacerbated by anecdotes coming from the business community. As a result, during the second half of the 1980s U.S.-Japan relations began to break down. It is at this time when trade friction noticeably intensified, when political restraint on Japan became formalized in legislation and bilateral agreements, and when 'Japan bashing' became an acceptable posture in Washington. The fact that U.S.-Soviet relations began to improve at this time created additional problems in that national security matters for the United States became more noticeably related to economic problems.

The collapse of the Soviet Union eliminated the opposing political corners existing in the international ring that had staged the conflicts of the Cold War. By the early 1990s relations between Japan and the United States had to be viewed without the prism that had been used to analyze Cold War politics. Indeed, in the 1990s ideological conflict has been replaced by inter-capitalist economic conflict. The counterpart to America's Japan bashing is very evident in Japan. *Kenbei* (the dislike of America) has become a fact of life in modern Japan. American policy makers continued use of tough talk when dealing with bilateral trade problems arguably have had the recent effect of creating a noticeable decline in the large amount of trust the Japanese once had of the United States.[16]

The unfair trade issue

Although U.S. policy makers were willing to ignore some economic pain in trade relations with Japan in exchange for a permanent military presence in the country,[17] economic and political thresholds were surpassed during the 1980s. The accelerated pace of Japan's industrial competence that led to rapid export growth in manufacturing together with the weakening of many American industries, including some high tech ones, signaled the need for aggressive political action to correct the growing U.S. trade imbalance with Japan. Loudly claiming unfair trade and attempting to mitigate the trade problem, many U.S. policy makers and analysts saw protectionism in the form of managed trade as the best way to handle Japan.

While the United States continued to rebuke Japan about the unfairness of its aggressive export and industrial policies, and for selecting winners,[18] most scholarly work in recent years has turned to the issue of closed Japanese markets. The evident U.S. concern with alleged obstacles standing in the way of the free movement of American exports to Japan is ironic. For decades, the United States has limited sectors of its domestic market to Japanese producers. It is to these issues that we now turn.

Japanese exports to the United States

Not long after U.S. policy makers became concerned about Japanese textile imports in the early 1950s, voluntary export quotas came into existence. In 1955 the United States negotiated the first voluntary quota with Japan on cotton blouses. Six years later 25 Japanese products were being controlled by voluntary export quotas with the United States. At this time somewhere between 25 and 35 percent of Japanese exports to the United States were subject to export quotas.[19] Three decades later this form of trade discrimination was still evident. In 1991 approximately 30 percent of Japan's merchandise exports to the United States were controlled by voluntary restraint agreements, thus going a long way in protecting the U.S. market and helping to manage bilateral trade.[20] While voluntary restraints are designed to be 'temporary,' some have lasted for many years: for example, restraints on Japanese steel remained in effect for over 20 years, those on automobiles continued for more than a decade, and those on machine tools, originally agreed to last for 5 years, were stretched to 7 years.

As explained in detail in Chapter 7, in recent years Japan has become increasingly vocal about how it has not been treated fairly by the United

States in trade matters. This includes the unilateral steps that have been taken by the United States.[21] Perhaps the best example of this is Super 301, a policy tool giving the president the authority to identify a nation as being in violation of U.S. law and accordingly permits the imposition of sanctions. Japan has also gone to great lengths to stress the very liberal use of anti-dumping measures by the United States. For now we need only to point out that while the Clinton administration's Japan policy stresses that Japanese markets continue to be closed to foreign producers, as of December 1995 the United States had over twice as many anti-dumping measures imposed on Japan as the combined total for Canada and the entire European Community.[22]

Japan has argued that the continuation of anti-dumping measures and the inconsistent use of 'rules of origin,' both of which are discriminatory, have offset America's much-reduced reliance on voluntary export restraints in recent years. By increasing the price of imported products, the discretionary use of rules of origin by a nation can simultaneously drive up investment costs for foreign transplant companies, specifically Japanese firms in the United States. According to the 1995 unfair trade report published by the government of Japan, the arbitrary use of rules of origin by the United States is discriminating against Japanese products because they have not been neutrally applied but rather have been used as a trade policy tool. Specifically, the Japanese cite the example of EPROM (erasable, programmable, read-only memory) in which U.S. rules of origin increased the number of products included in an anti-dumping investigation which ultimately was unfavorable to Japan.[23]

An interesting feature of America's efforts to manage Japanese imports is that the products identified as being 'dumped' in the United States have relied increasingly on larger and larger amounts of technology. In the 1950s and 1960s the United States managed restrictions on Japanese textile products; subsequently, steel, TVs, automobiles, video tape recorders, machine tools, flat panel displays, and semiconductors faced restrictions. It is this steady progression of S&T competency that is commonly left out of U.S. explanations of alleged dumping by Japan. This competency component is critical since Japanese goods have often provided American consumers with a better (and sometimes just about the only) product choice (e.g., flat panel displays and video cassette recorders).

Moreover, the cost component of products alleged to have been dumped by Japan in the United States needs to be seen in a context different from that used by American analysts. While some U.S. industries (most notably aircraft and semiconductors) first made use of 'forward pricing,' American trade representatives have disregarded this practice when calculating the price structure of several products, particularly advanced technology merchandise, alleged to have been dumped by Japanese businesses. In using forward pricing a business makes the reasonable assumption that as it discovers how to better manufacture a product and progresses up the learning curve, it will realize greater manufacturing and technology efficiency. In short, costs will decline as research and development and the improved efficiency from capital investments are better integrated into the production process. In the meantime, the product will temporarily be sold below higher short-term costs.[24]

Thus while it is often very difficult to prove unequivocally that dumping has occurred, when forward pricing is used, as it commonly is on advanced technology products, the job becomes even harder. There is, in other words, a fairly substantial gray area where uncertainty exists: are manufacturers dumping products to gain market share unfairly *or* are they simply making use of past experience and anticipating declining costs over time? Add to this the point that dumping has become commonly associated with Japanese trade behavior, so either many Japanese exporters have been inclined to dump their products, or the United States has too hastily responded that goods have been dumped when this is not so or where uncertainty exists. The position here is that dumping is not something that characterizes the trade practices of Japanese industries any more so than the businesses of other advanced nations.[25] However, politically it has been easy for the United States, which has dominated the post-war alliance, to convey the impression that Japan has dumped its products, notwithstanding the amount of this practiced by Japanese companies.

Closed Japanese Markets?

During the 1980s the United States became increasingly determined to have more access to the Japanese market in order to correct its growing trade imbalance with Japan. This determination was fully countenanced by official policy initiatives. Part of the American concern about the bilateral trade imbalance during the first half of the 1980s was the low value of the yen relative to the dollar. In 1984 U.S. officials began pressuring Japan to continue deregulating its financial markets and also to internationalize the yen. In both cases, U.S. policy makers wanted to drive up the relative value of the yen against the dollar. It was thought that this ultimately would improve the bilateral trade imbalance by making American products more appealing to Japanese consumers.

More directly related to product areas, in 1985 the United States initiated the MOSS (market-oriented sector-specific) talks with Japan. The MOSS discussions centered on four sectorial areas -- medical equipment and drugs, forest products, telecommunications, and advanced electronics -- and involved an array of access impeding problems identified by the United States. Bilateral discussions on semiconductors (which were not dealt with extensively in the MOSS talks) also emerged during the 1980s, with the United States wanting Japan to import more American chips. Finally, in 1989 the United States persuaded Japan to take part in the Structural Impediments Initiative, with the explicit intention of increasing American exports to the Japanese market.[26]

It must be understood that short of implementing measures that would strengthen America's industrial and S&T base, managing trade with Japan by concentrating only on guarding the U.S. market would have been politically and economically unwise. Demanding only that Japan voluntarily reduce its exports to the United States in conjunction with implementing and sustaining numerous anti-dumping measures would have intensified the suspicion that the consumer attractiveness of products made by U.S. businesses had declined to the point where government had to take action and protect weakened domestic industries. Partly for this reason, more attention had to be given to the alleged closedness of Japan's domestic markets. Because Americans had discovered that Japanese products were a preferable substitute for U.S. merchandise, other than protecting the domestic market, there was no other immediate option available that could bring relief to the trade imbalance. Thus by increasing U.S. exports to Japan, the United States could reduce its trade deficit with Japan. Also important was that the claim of closed markets could easily be linked to job loss in the United States, creating public support for the get-tough approach toward Japan.

During the summer of 1981 policy makers within the Reagan administration had given strong indication that their public efforts would mainly zero in on the political issue of opening up Japan's domestic markets. Were Japan's tariffs really prohibitive to foreign producers? This was not the case in the early 1980s, and it is not true now. While it is true that Tokyo had overprotected its domestic industries in the past to help develop and sustain economic growth, Japan's average tariff rates for many years have been lower than those for the United States, Canada, and the

European Community. Moreover, Japan's nontariff trade barriers on imported manufactured products are below those maintained by most other OECD (Organization for Economic Co-operation and Development) members.[27]

For more than a decade, there have been empirical studies performed by analysts in the United States attempting to assess the extent of openness in the Japanese domestic market. There is, however, no consensus on this matter. Some studies suggest that Japan is an 'outlier' (that is, its domestic market is comparatively closed) among industrialized capitalist nations, with relatively low intra-industry trade.[28] Other studies suggest that Japan is really not that different.[29] Although the academic debate continues in the United States, the OECD has stressed the inconclusiveness of this matter and points out that Japan's undivided attention to technological enhancement may very well have provided it with an edge in establishing comparative advantages that help to account for its relatively low intra-industry trade.[30] A recent empirical work performed by a Japanese government researcher, moreover, significantly reduces the plausibility of the leading American studies that attempt to show that Japan's domestic markets are closed to foreign products by calling attention to the existing deficiencies in the explanations.[31] There is, in short, no compelling scientific evidence that demonstrates conclusively that Japan's domestic markets are closed to foreign producers.[32]

The failure of empirical studies to prove convincingly that Japan's domestic markets are closed to foreign producers of manufactured goods is interesting. For more than a decade, both republican and democratic administrations in Washington have built public relations campaigns on the assumption that Japan's domestic markets are closed to American manufactured products. However, reality suggests a different picture, or at least suggests that the one U.S. policy makers display publicly may not be authentic. Thus with regard to the issue of Japan's alleged closed domestic market, U.S. policy has for years rested on uncertain empirical ground, filled in with 'maybe' and 'appears to be' statements.

Moreover, the Japanese study referred to above raises a very important point. Although not fully articulated, the study suggests that in one U.S. econometric model it is unclear whether a measurement estimates Japan's closed domestic market or the closedness of foreign markets to Japan's exports. We have already pointed out that the United States has freely employed anti-dumping measures against Japan. Add to this those imposed by the European Community, Canada, and elsewhere and we find that

Japanese products confront a large number of anti-dumping measures -- a total of 79 as of December 1995, many of which have been in effect for years. Voluntary export agreements, as already noted, have also reduced exports from Japan to the United States. Thus a reasonable argument is that confronted with obstacles that retard exports, Japanese producers have compensated for this problem by concentrating on strengthening domestic consumption for their products.[33]

In any case, the shortcomings of empirical studies, i.e., their not showing the closedness of Japan's domestic markets, have been politically offset by anecdotes from U.S. businesses. Complaints by U.S. business people concerning the alleged closedness of Japan's markets have circulated in Washington for years. Story after story has been told of the difficulties faced by U.S. business people attempting to do business in Japan.

What stories are not heard are those that explain the successes of U.S. business people in Japan, which should become even more plentiful because of recent initiatives by the Japanese government designed to facilitate foreign investment.[34] In one statement made by a very grateful sales representative of American Ergonomics Corporation to JETRO (Japan External Trade Organization) it is made clear how extensively this organization worked to help his firm establish business relations in Japan.[35] In another statement by an American citizen who has lived and worked in Japan for more than two decades and who is the president of K.K. Ashisuto, a Japanese computer software firm, we are told that Japan is a difficult place to do business only if a company prefers not to make adjustments. It is obvious that his firm did make the adjustments, since its computer software sales exceeded 9 billion yen ($65.2 million) in 1989, and controlled almost a quarter of the Japanese market; interestingly, 95 per cent of the personal computer software products sold by Ashisuto in Japan were imported from the United States.[36] He goes on to point out that it has been his willingness and determination to learn what is important in Japan, including the business relations, that has made his firm successful. He writes:

> There are Japanese losers as well as foreign losers. We hear more from the losers because they cry and scream so loud. Or get their politicians to cry and scream for them. The winners are usually quiet because they have nothing to gain by telling their competitors about the great opportunities in Japan or teaching their competitors how to realize those opportunities. Does Macy tell Gimbel? Why should Coke tell Pepsi? What has BMW to

gain by instructing Chrysler?

Indeed, unsuccessful Chrysler Corporation has complained loudly. Former Chrysler chairman of the Board of Directors, Lee Iacocca, claimed in a speech given to Detroit's Economic Club soon after returning from a visit to Japan with former president Bush and other U.S. business people that its domestic market is a 'sanctuary' and that the Japanese 'won't allow any foreign competitor to be a factor in their market.'[37]

As Table 4.1 indicates, there are a number of American businesses that have been successful in Japan. Some other U.S. companies that have done well in Japan are Amway, General Electric, Caterpillar, and Dow Chemical. Generating the sales of American firms operating in Japan are imports as well as goods manufactured there.[38]

The most obvious question here is how can some American firms succeed in the Japanese market if it is really closed to foreign business? Success, whether it be in Japan or in any other industrial nation, depends upon effort. During recent years, according to a study supported by the National Science Foundation, more than 200 American companies have been developing plans to begin doing research and development work in Japan.[39] A number of American firms determined to make the Japanese market work for them have already established wholly owned or joint venture research and development facilities in Japan. As of the early 1990s, there were over seventy American companies that had selected this course for the purpose of expanding their business operations in Japan. The use of the word 'expanding' is not haphazard, since commonly companies have initially established sales offices and then later invested in production and research and development operations so that they can better adjust their efforts to the Japanese market.[40] W. R. Grace and Company is somewhat typical of this pattern. Having first established an office in Japan in 1960, W. R. Grace had made the decision by the mid 1980s to open a research and development facility there to enhance its business prospects.[41] It needs to be understood that a closed market would not tolerate this kind of technological penetration by foreign business. Moreover, an analysis of 50 nations' political and business conditions by a Swiss firm indicates that Japan is one of the best places (noticeably ahead of the United States) for new foreign businesses to earn profits; this, indeed, is not at all compatible with the argument that Japan's domestic markets are closed to foreigners.[42]

Ultimately, however, just how open or closed Japan's markets are to

Table 4.1 American businesses in Japan

1. Fuji Xerox Company, Ltd: 50 percent owned by Rank Xerox Ltd., a subsidiary of Xerox Corporation; 1990 sales in Japan, 539.2 billion yen ($3.7 billion); held 13.8 percent of Japanese copy machine market.

2. NCR Japan, Ltd.: 70 percent owned by NCR Corporation, which is part of the AT&T organization; 1990 sales in Japan, 119.4 billion yen ($823 million); held 28.5 percent of the Japanese terminal market.

3. Yokogawa-Hewlett-Packard, Ltd.: joint venture, 75 percent owned by Hewlett-Packard Company; 1990 sales in Japan, 160.1 billion yen ($1.1 billion).

4. IBM Japan, Ltd.: owned entirely by IBM World Trade Corporation; 1990 sales in Japan, over 1.326 trillion yen ($9.1 billion).

5. Texas Instruments Japan Ltd.: owned entirely by Texas Instruments, Inc.; 1990 sales in Japan, 149 billion yen ($1 billion), with semiconductors making up 80 to 90 percent of sales; company held 2.7 percent of the Japanese chip market.

6. Nihon Unysis, Ltd.: joint venture with Unysis; 1990 sales, 285.8 billion yen ($2 billion).

7. Nihon Digital Equipment Corporation: owned entirely by Digital Equipment Corporation International; sales in 1990, 128.7 billion yen ($888 million).

8. Pfizer Pharmaceuticals Inc.: owned entirely by Pfizer Oral Car Corporation, a subsidiary of U.S. Pfizer Inc.; 1991 sales in Japan 78.5 billion yen ($583 million).

9. Esso Sekiyu: owned entirely by Esso Eastern, Inc.: 1990 sales in Japan, 613.6 billion yen ($4.2 billion); held 6.4 percent of Japanese gasoline market.

10. Mobile Sekiyu: owned entirely by U.S. Mobile Petroleum Company, Inc.; 1990 sales in Japan, over 696.2 billion yen ($4.8 billion); held 7.5

percent of Japanese gasoline market.

11. McDonald's Company (Japan) Ltd.: 50 owned by McDonald's Corporation; 1990 sales in Japan 175.5 billion yen ($1.2 billion); held an estimated 40 percent of Japanese hamburger market.

12. Ajinomoto General Foods Inc.: 50 percent owned by U.S. Kraft General Foods; 1990 sales in Japan, 79.5 billion yen ($548 million); held 25 percent of instant coffee market in Japan.

13. Coca-Cola (Japan) Inc.: entirely owned by Coca-Cola Export Corporation; 1990 sales in Japan, 175.6 billion yen ($1.2 billion; held 92 percent of Japanese cola market.

14. Levi Strauss Japan: 83.6 percent owned by Levi Strauss & Company; 1990 sales in Japan, over 28.9 billion yen ($199 million); held 9.2 percent of blue jean market in Japan.

15. Cray Research Ltd.: by the end of 1991 Cray had sold 39 supercomputers in Japan.

16. Nippon Motorola Ltd.: owned entirely by Motorola Inc.; company has experienced over 40 percent growth per year in Japan.

17. Apple Computer Japan: owned entirely by Apple Computer, Inc.; company has experienced steady annual growth of personal computers and related items in Japan.

18. Nihon Sun Microsystems: owned entirely by Sun Microsystems, Inc.; company experienced steady growth of sales in Japan, recorded a 65.9 billion yen ($454 million) growth in sales in 1990.

19. Du Pont Japan Limited: owned entirely by E.I. Du Pont de Nemours & Company (U.S.); sales in Japan have grown steadily and increased by 102 billion yen ($703 million) in 1990; held 20.4 percent of graphite, polyethylene, and terephthalate resin market.

20. Proctor and Gamble Far East Inc.: firm held 12 percent of market share for household powered detergents, 23 percent of sanitary napkin market, and

26 percent of the disposable diaper market.

21. Kodak Japan Ltd.: 51 percent owned by Eastman Kodak Company; held 12 percent of Japanese color photographic film (general use) market in 1990.

22. Warner, UIP: UIP held 27.1 percent and Warner held 9.6 percent of the 71.1 billion yen ($490 million) worth of movie sales in Japan in 1990.

Source: Ministry of International Trade and Industry, Challenges to the Japanese Market, Ministry of International Trade and Industry Overseas Publication Office, January 1992.

foreign business really reduces to little more than opinion. Because there is no definitive indicator determining the degree of market accessibility by foreign business, allegations of closed markets could therefore be directed at any nation, including the United States. Given the restrictions on Japan's exports to the United States, the Japanese have found it easy to make the argument that America's markets have not been fully accessible to their products.[43]

When looking at imports as a percent of manufacturing production over a number of years, the difference between the United States and Japan is not that great. Table 4.2 shows that imports as a share of total manufacturing production have increased significantly for both Japan and the United States. It also shows that the difference between these two nations is considerably smaller than the impression given by politicians and others claiming that Japan's domestic markets are closed to foreign business.

Moreover, after examining the import share held by foreign high technology producers in the domestic markets of the United States, Japan, Germany, France and Italy, one could easily conclude that *both* the United States and Japan are 'outlier' nations. In 1992, for example, the import market share held by foreign producers was 28.5 percent in the United States, 18.7 percent in Japan, 67.4 percent in Germany, 77.7 percent in France, 81.5 percent in the United Kingdom, and 66.4 percent in Italy. One could also conclude that specific U.S. high tech markets are largely closed to foreign manufacturers. For example, imports accounted for only 6.3 percent of the drugs and medicines market in the United States, and for only 8.9 percent of the aircraft industry market.[44] Thus, if Japan's high technology markets are closed to foreign business, the data suggest that some of America's

Table 4.2 Ratio of imports to production, manufacturing industries, 1981-1992

Year	United States	Japan	Difference
1981	9.4%	5.4%	4.0%
1982	9.4	5.6	3.8
1983	10.5	6.0	4.5
1984	12.8	6.7	6.1
1985	13.9	6.8	7.1
1986	15.7	9.6	6.1
1987	15.9	12.0	3.9
1988	16.0	13.7	2.3
1989	16.9	14.9	2.0
1990	17.2	14.9	2.3
1991	17.6	14.9	2.7
1992	18.6	15.8	2.8

Source: Calculated from National Science Board, Science & Engineering Indicators - 1993 (Washington: D.C.: United States Government Printing Office, 1993), p. 440.

markets are not much more open.

If Japanese markets are closed to foreign businesses, then one must assume that *gaiatsu* (external pressure) originating from the United States completely explains why America's merchandise exports to Japan have increased so dramatically for more than a decade in real dollar value. (See Figure 4.1.) One would largely have to ignore the independent effect of the very significant improvement in Japan's per capita gross domestic product and the phenomenal growth of industry over the past few decades. Would not these developments create some additional demand for foreign goods, including the consumption of products that incorporate advanced technologies? The data suggest that the answer is yes.

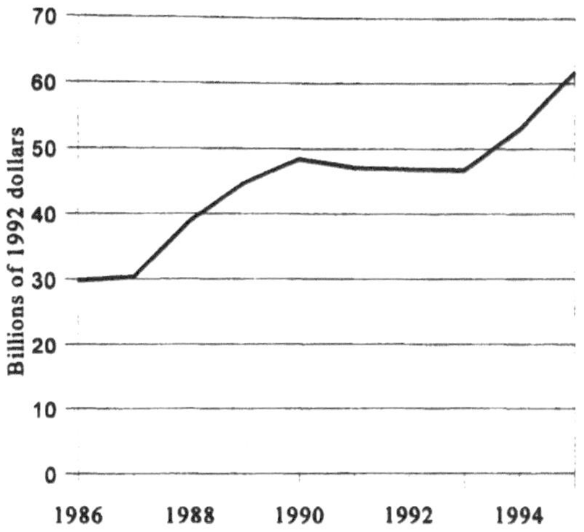

Figure 4.1 Real U.S. merchandise exports to Japan 1986-1995

Sources: Economic Report of the President (1996); and Office of U.S. Trade Representative, 1996 National Trade Estimate.

Table 4.3 shows that real high tech exports from the United States to Japan nearly quadrupled between the late 1970s and 1990. Table 4.3 also shows that as a percentage of total U.S. high technology exports, the proportion going to Japan has grown markedly since the late 1970s. Indeed, the dollar value of U.S. high technology exports to Japan has exceeded that to France, Germany, and the United Kingdom respectively since 1979. Between 1984 and 1993 the dollar value of several U.S. high technology products exported to Japan have shown appreciable growth: computers by 315 percent, computer parts and office machines by 244 percent, communications equipment by 483 percent, aircraft and parts by 262 percent.[45]

This aside, it has been argued that Japanese businesses in the United States export a disproportionately large amount of goods back to Japan. According to this line of reasoning, this accounts for a substantial part of the observed export growth from the United States to Japan.[46] However, this is only true for merchandise exports (a generally inclusive category) from Japanese businesses operating in the United States, and there is a reasonable explanation for this. Relative to other industrial nations, some intra-firm merchandise exports from Japanese affiliates operating in the United States to their parent companies in Japan are noticeably higher than other industrial

Table 4.3 Real U.S. high technology exports to Japan, 1978-1990

Year	Billions of (1987) $	% of Total high tech
1978	$3.5	6.4%
1979	4.6	7.8
1980	5.0	7.6
1981	5.7	8.2
1982	5.1	8.0
1983	6.2	9.6
1984	6.0	9.2
1985	7.0	9.9
1986	8.3	11.2
1987	8.8	10.4
1988	10.4	10.8
1989	11.0	11.0
1990	12.6	11.2

Sources: Calculated from National Science Foundation Data obtained from Data Resources Inc./McGraw-Hill; and Economic Report of the President, (1994).

nations. Because Japan is not a resource rich nation and similarly lacks other basic goods needed to sustain itself, it must rely on imports. Thus in a list of 11 such products (e.g., food and crude materials) exported by foreign affiliates operating in the United States, Japanese affiliates held the first position in 8 of them. Most significant, Japanese affiliates in the United States accounted for more than 80 percent of all foreign affiliate exports of petroleum, nearly 60 percent of crude materials, and almost half of all food.[47]

An examination of the manufactured exports of Japanese affiliates operating in the United States, however, sheds a different light on our analysis, since in recent years they have been more or less proportionally the same as several other industrial nations. For example, in 1991 the proportion of all foreign affiliate manufactured exports from the United States accounted for by Japanese businesses was 15.4 percent, compared to 14 percent for Canada, 14.1 percent for France, 14.8 percent for Germany, and 15.7 percent for the United Kingdom. In fact, in the past the Japanese share was actually lower than some other industrial nations. In 1985 the Japanese affiliate share was 6.6 percent, compared to 16.2 for the United Kingdom, 14.1 for Germany, and 26.2 percent for Canada.[48]

While Figure 4.2 demonstrates that real U.S. manufactured exports to Japan have grown sharply over the years, Figure 4.3 is even more telling. Figure 4.3 shows that after taking out exports to Japan originating from Japanese affiliates operating in the United States, American manufacturing exports to Japan have still grown very significantly, nearly doubling in real dollar value between 1985 and 1991. Despite whatever amount of weight one may attach to yen appreciation relative to the dollar, giving Japanese buyers more foreign purchasing power in recent years, the fact remains that more goods originating from nonJapanese-affiliated businesses have been exported from the United States to Japan. Thus it is entirely reasonable to conclude that the noticeable growth of manufactured goods exported from the United States to Japan by nonJapanese affiliated firms indicates a market that at the very least is moving more toward openness than remaining closed to foreign products.

One should not be led to believe that the 'results-oriented policy' of the Clinton administration has caused more U.S. exports to Japan. As we have seen here, U.S. exports to Japan were growing well before the Clinton administration took charge. That the ratio of Japanese imports of American manufactured goods to total imports from the United States -- which includes Japan's voluminous agricultural imports -- increased from 44.1 percent in 1980, to 62 percent in 1990, to 66.4 percent in 1995 is indicative of a trend that antedates the Clinton administration.[49]

Our discussion on U.S. exports to Japan raises an important question: what amount of American exports to Japan determines when Japanese markets are sufficiently open? Without knowledge of this and since managed trade in a bilateral relationship is fundamentally a power struggle, pressure by the United States on Japan to open its markets could continue indefinitely. A good example of this is semiconductors. In the 1980s when the semiconductor dispute appeared U.S. complaints of unfair trade included both charges of dumping and insufficient access to Japan's domestic market. For these reasons, the U.S.-Japan semiconductor agreement signed in July

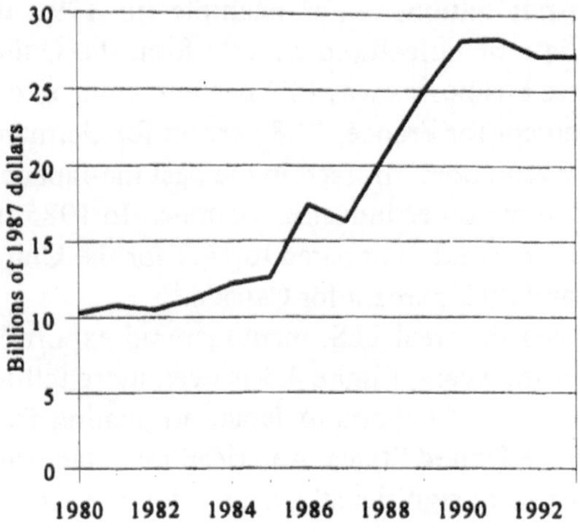

Figure 4.2 Real U.S. manufacturing exports to Japan, 1980-1993

Source: U.S. Department of Commerce, various publications.

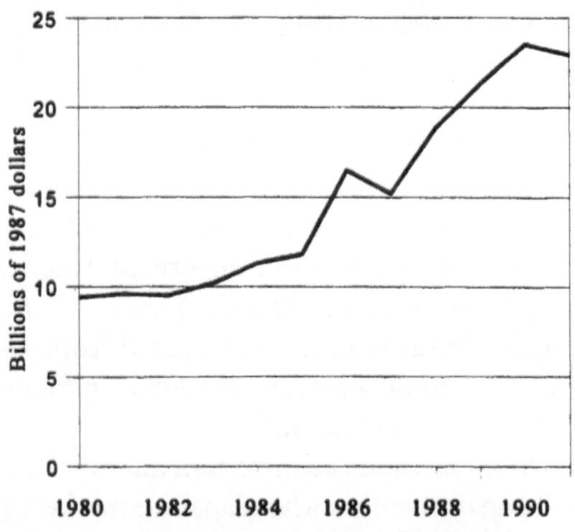

Figure 4.3 Real U.S. manufacturing exports to Japan excluding Japanese affiliates in the U.S., 1980-1991

Sources: Ibid.; and U.S. Department of Commerce, Survey of Current Business (William Zeile), October 1993.

1986 both protected the domestic market for American manufacturers and negotiated a predetermined (20 percent) portion of Japan's market that would be held by foreign producers.[50] Because the 20 percent target had not been reached by 1991, this precipitated the implementation of a new semiconductor agreement which again established a quantitative proportion of Japan's market to be held by foreign producers. By the final quarter of 1992 this 20 percent target had been slightly surpassed (20.2 percent), although all of 1992 was less satisfactory to the United States.

The Japanese felt it important to point out in early 1993 that the failure of U.S. companies to sell more semiconductor chips in Japan during 1992 was largely their own doing. Increased demand for chips in the United States during 1992, according to Japanese government and business officials, caused American firms to cancel, reduce, or delay the delivery of chips to Japan. MITI announced in early 1993 that it had determined from a survey by the Electronic Industries Association of Japan that an estimated $60 million worth of chip sales to Japan had been canceled by U.S. firms, and pointed out that the actual amount could be much greater.[51] The waters got rougher in early 1993 when the foreign share of Japan's market declined slightly below the benchmark. This prompted the Clinton administration to apply renewed pressure on Japan to maintain the 20 percent minimum share.

Meanwhile, the U.S. Semiconductor Industry Association lost no time. By March 1994 it was insisting that the foreign share of Japan's semiconductor market should be increased and fall within a range of 25 to 28 percent by 1996.[52] Since American producers held a considerably higher share of the European market (about 60 percent in the late 1980s),[53] semiconductor industry officials and U.S. policy makers had available a political justification for insisting that the foreign share of Japan's domestic market be increased in the coming years.

However, it must be stressed that Japanese semiconductor producers are amongst the best and most efficient in the world, and differ substantially from their less accomplished European counterparts. Thus the existence of a quantitative target based at least in part on a comparison with the European market was a completely unreasonable way to expect semiconductor trade to be conducted between Japan and the United States.

While the request of the U.S. Semiconductor Industry Association never materialized, toward the end of the 1991 semiconductor accord, which expired at the end of July 1996, the foreign share had reached 30 percent of the Japanese semiconductor market. The United States still insisted that the bilateral semiconductor agreement should be extended beyond the July 1996 deadline, despite the appreciable share of Japan's domestic market held by foreign producers.[54]

Because there is no indicator that reflects market closedness and since in a managed trade relationship power is often the deciding criterion, Japan continues to contend with the demands and complaints of the United States. Indeed, growing exports and a declining trade deficit have not ended American complaints from both government and the private sector that Japan's domestic markets are still closed. For example, in obvious disagreement with his Japanese counterpart, the American chairman of the Japan-U.S. Business Conference stated at the organization's Tokyo meetings in July 1996 that: 'We are making some progress. Still, serious problems remain.'[55]

We must finally address the question of why the issue of an overly protected Japanese market has become a source of U.S. complaints only since the early 1980s. With tariffs in place protecting many industries, Japan's domestic market was much less accessible to foreign manufacturers in the 1960s and 1970s than it has been in more recent years. For this reason, the inaccessibility of Japan's domestic market is not really the central issue. Rather, the point is that over the years Japan has markedly increased its manufactured, including high technology, exports to the United States (and to other nations), giving it a relatively big trade surplus. It is this condition combined with Japan's subordinate position in its post-war alliance with the United States that largely explains the continued onslaught of complaints relating to closed markets and unfair trade generally.

Industrial groups: the *keiretsu*

One of the most critical issues involved in the ongoing debate concerning the claims of Japan's closed markets is the *keiretsu*, or industry organizations. Emerging after World War II, the *keiretsu* have a long and important economic history in Japan.[56] Some have direct linkages to the *zaibatsu*, which were holding companies that existed in Japan until the end of the war and were dissolved by the American occupation.

The *keiretsu* evolved in Japan to facilitate commercial activity and its longevity indicates a continuing correspondence with existing conditions. It did not develop as a business subterfuge designed to 'seal off' Japan's domestic market from foreign business. Nonetheless, many U.S. observers have maintained that despite the declining presence of formal trade barriers, informal obstacles such as the *keiretsu*, still severely limit access to Japan's domestic market by foreign manufacturers.

There are two major types of *keiretsu*: the vertical and the conglomerate or intermarket types. As they work within the Japanese economy these two types of industrial organizations coincide to some extent with each other so a firm may be part of more than one *keiretsu* organization. Major characteristics of the *keiretsu* include financial support to members, common stock ownership, and, significantly, long-term business relations that have produced both mutual trust and commitment to the organization. This commitment includes that of workers to companies. Although said to have been endangered by the recession in the first part of the 1990s, lifetime employment which has roots that go back more than seventy-five years and overlaps with seniority wages still exists in Japan, especially in big firms that are part of a *keiretsu*.[57]

The existence of the *keiretsu*, in particular the vertical type, is said to account for why intra-industry trade in Japan is lower than it is in other industrial nations. But intra-industry trade for the six of the largest corporate groups stood only at 8 percent of their total procurement in 1992. Although a somewhat larger percentage existed in the past, intra-industry trade was even then only a small percentage of *keiretsu* business activity.[58]

The Japanese have an entirely different view of the *keiretsu* than most American analysts. As they see it, the *keiretsu* is dissimilar to the *zaibatsu*, since the latter was the monopolization of the Japanese economy by a small number of families, and after 1931 often closely related to the militarists in Japan; one could hardly make this claim about the *keiretsu*. The Japanese also point out that the *keiretsu* system is not purposively designed to manipulate markets or to exclude foreigners from active participation in Japan's domestic economy. Moreover, some U.S. firms have accessed the *keiretsu* 'on the fringes,' that is, through sales networking and joint ventures, while others such as IBM and Coca Cola have challenged them head on.[59]

The main purpose of the *keiretsu*, say the Japanese, is to ensure against unfriendly takeover attempts. The Japanese argue that the *keiretsu* should rightly be compared to the big fully integrated U.S. or other Western company that literally does most of the manufacturing work itself and which, as much as possible, seeks to avoid providing contract work to outside firms, such as in the automobile industry. Increased technological and marketing efficiency, say the Japanese, result from the *keiretsu* system. In brief, the Japanese argument is that it is wrong for Westerners to criticize the *keiretsu* system simply because of its organizational complexity.[60]

Technically the *keiretsu* is the adaptation of an institutional mechanism to Japan's business environment. Because its development is considerably more of a natural rather than a deceptive expression of Japan's business environment, it is therefore wrong to allege unfairness based on the existence of this institutional mechanism. To understand why Japan's home market came to rely on the products of domestic manufacturers to a somewhat greater extent than what has existed in other industrial nations, it is necessary to look beyond the *keiretsu*. To do this, we must examine several important political, geographic, economic, and cultural factors that converged to create this condition.

First, to reduce the external influence that would thwart the efforts to rebuild Japan's ruined economy, government officials passed laws with U.S. approval shortly after World War II. Laws such as the Foreign Exchange and Foreign Trade Control Law effectively protected the domestic market for some time, and consequently encouraged close business ties within industry groups as well as promoting the widespread availability of Japanese-made products. Although Japan formally eliminated the restrictive nature of most of these laws years ago,[61] established relations and consumption preferences could unwittingly favor the *status quo*. It should be understood, however, that these behaviors are somewhat analogous to the 'buy American' mentality that has existed in the United States, even though it is true that this sentiment has been shattered by the appearance of quality foreign products.

Second, the growth path selected by Japan has focused on some economic priorities that differ from those chosen by the United States. Encouraged by the U.S. occupation leadership, Japan legally committed itself to abandon any interest in an aggressive military; at the same time, this restriction helped to create conditions which left Japan with only one viable option. Because of the prohibition barring the development of a big and aggressive military and given the determination to rebuild the economy, Japan had to depend on civilian investment, to which government guidance became an important part. In addition to expanding exports, the development of manufacturing competence permitted Japanese businesses to serve their domestic market. Associated with this is that Japanese manufacturers have given much attention to technology development for the purpose of providing continuous product improvement.[62]

A third significant factor that has reduced the foreign penetration of Japan's domestic market has been its location. Unlike Western European nations and even the United States, geographically Japan is a relatively isolated nation. Finally, because the Japanese are a very traditional people, mutual respect, caring about others, and commitment and loyalty to the group have long been a part of their way of life. The fact that over 99 percent of the people of Japan are of Japanese descent undeniably creates a bonding and a sense a national loyalty that to some extent contributes to a predilection to purchase goods produced by domestic producers.

Along with the possibility of discrimination against Japanese goods in foreign markets which forces more aggressive domestic marketing, together all of the above matters produce a consumption momentum that has encouraged a preference for Japanese goods. It must be stressed, however, that this is not the same thing as saying that Japan's domestic market is closed to foreign business today. What these combined factors do encourage is a consumption priority that is based on the actual and potential benefits of maintaining established sociocultural relations. This means that it is the responsibility of foreign businesses that want to succeed in Japan to work the Japanese market and the supporting institutional network -- for example, JETRO -- to establish business relations.

As we have suggested above, foreign businesses wanting to do well in Japan must be committed to make the necessary adjustments that will produce the desired results. This is not something too many American companies have been willing to do. U.S. automobile manufactures, for example, have wanted to sell cars in Japan but have not been willing to make the necessary and required adjustments to prevailing conditions there. The same is true for American pharmaceutical companies.[63] Some American companies wanting to sell in Japan have simply attempted to transfer business practices or products from the United States to Japan. Much to their dismay, they have learned that results have not been as good as they expected, and this is because they have not been willing to make the necessary commitment to the Japanese market.

Despite this resistance evidenced by some American companies, Japan has indicated that it is willing to assist American business. In 1994 Japan's Export-Import Bank entered into negotiations with Ford, offering to make a substantial low-interest (about 4 percent) loan in the range of $200 million to $300 million to help the company produce cars with right-hand steering which would be exported to Japan. In fact, Chrysler and General Motors have also spoken with Japan's Export-Import Bank about low-interest loans, although talks have not proceeded as far as they have with Ford. The Japanese, however, made it perfectly clear that the Ford loan is dependent upon the Clinton administration's backing off its position relating to demands that Japan accept a prearranged amount of U.S. goods.[64]

Although the Japanese government has tied a political rope to the availability of this loan, nonetheless the funds will permit a necessary adjustment and provide a sign of a nascent commitment to Japan's domestic market. A large part of a foreign company's demonstrated commitment to Japan's domestic market is its willingness to make available a 'quality' product to the Japanese. Indeed, a survey of foreign-affiliated companies operating in Japan by the Japanese Fair Trade Commission shows that the leading difficulty faced by these firms is 'quality,' followed by 'delivery time and methods,' and 'price.'[65] Given that Japanese firms have committed themselves to consistent quality improvements, the Japanese people have become increasingly accustomed to quality. They therefore expect quality when considering making purchases from foreign firms also.

The establishment of a research facility in Japan by a U.S. company suggests that it intends to make a 'quality investment' there and that it wants to explore product requirements and to make adjustments when necessary. A survey supported by the National Science Foundation found that the majority of American companies, nearly 90 percent, with a research facility in Japan are involved in product development work, that is, they are there to market goods, but only after adjusting them to the Japanese market when necessary. This same survey asked American companies if they were permitted to become part of Japanese research or industry associations. The overwhelming majority, 94 percent, responded that they were permitted to join.[66]

Thus quality and, again, a demonstrated commitment to the Japanese market are essential to operate successfully in Japan. U.S. companies doing business in Japan that fail to incorporate these ingredients into their plans are less likely to be successful than those that do employ them. When expectations of success diminish or are no longer present, identifying differences in business milieus becomes perhaps the most likely line of defense for American businesses. Because the *keiretsu* is easily identified as being different, through the use of anecdotes it has commonly been singled out by U.S. companies and government officials as creating obstacles to free market business activity in Japan.

Is there compelling statistical evidence that suggests that the *keiretsu* lowers intra-industry trade and reduces imports? Although the academic debate continues on this question, the fact of the matter is that there is no good evidence indicating that the *keiretsu* substantially inhibits free trade between nations.[67] The central issue in the debate involving the *keiretsu* (efficiency versus exclusion) therefore should be interpreted as more of a political matter than anything else. Because it has not been shown that Japan's domestic markets are closed and that the *keiretsu* are largely responsible for this, U.S. insistence that *keiretsu* relations be dissolved (e.g., the Structural Impediments Initiative talks demanded this[68]) has actually served to maintain trade friction in the bilateral relationship.

Interesting too is that although U.S. policy makers and industry representatives complain loudly about the *keiretsu* there is a definite realization of its importance to increased business efficiency. In fact, some U.S. companies have taken steps to implement a similar system. A study of the U.S. auto industry provides some indication that in recent years American car makers have been working to forge long-term contractual relations with their suppliers.[69]

In perspective

The argument in this chapter has been that Japan has been confronting what largely amounts to exaggerated charges of unfair trade made by the United States. The lack of convincing evidence that Japan's domestic merchandise and especially its manufacturing markets are closed to foreign products makes the unfairness argument lose much credibility. Even when looking at foreign procurement by the Japanese government, the argument of the inaccessibility of Japan's domestic markets is weakened by the facts. The foreign share of foreign products in procurement by the Japanese government in 1992 was 14.9 percent for Japan, more than twice as high as it was for the United States (7.1 percent).[70] Moreover, to the extent that Japan has been practicing unfair trade, the United States (or most any other nation) could also be identified as participating to a similar degree in this kind of business behavior. Reducing the access of Japanese products to the United States through the widespread and protracted use of anti-dumping measures and by negotiating voluntary restraints on foreign exports, most of which had been in effect until very recently,[71] is not the policy approach of a nation practicing fair trade.

That some business behavior in Japan is different from that performed by U.S. companies does not demonstrate that its markets are closed or that it is unduly practicing unfair trade. The continued existence of the *keiretsu* similarly does not provide evidence of exclusionary or unreasonable

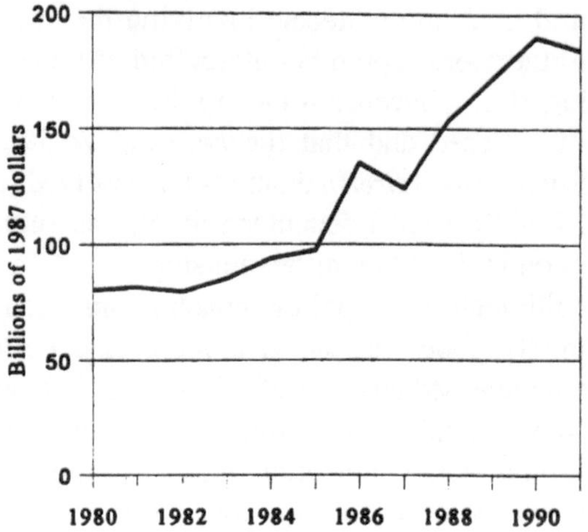

Figure 4.4 Per capita U.S. manufactured exports of nonJapanese-affiliated firms to Japan, 1980-1991

Sources: Same as Figure 4.3; and STA, White Paper on S&T, 1992.

discriminatory business practices. Its existence merely indicates a traditional disposition for using and maintaining organized business networks which, in contrast, are not as well established in the United States.

The closed market hypothesis cannot explain why the United States was the leading exporter of high technology products (OECD definition) until 1982, but subsequently fell behind Japan; nor does it explain why for years the ratio of high technology production to total manufacturing output in Japan has remained higher than that for the United States. This hypothesis furthermore cannot explain why the United States slipped behind Japan in total manufactured exports in the early 1980s. That the United States has historically been behind Japan and other industrial economies in the proportion of manufacturing output that it exports, however, does suggest something very important to our analysis: the continued insistence that Japan purchase more American-made products, which overlaps with the argument that its domestic market is relatively closed, is a political tactic that can help offset the voluminous Japanese manufactured exports to the United States.[72]

The fact that manufactured products exported to Japan by nonJapanese-affiliated producers in the United States have increased sharply over the years, as seen above, is compelling evidence available on the matter of the degree of openness of Japan's domestic market. Figure 4.4

demonstrates the striking growth of real manufactured exports from the United States by nonJapanese-affiliated businesses measured on a per capita basis, suggesting the increasing consumption of American-made products by the Japanese. It is hardly likely that a closed Japanese market would yield an increase of almost 130 percent in manufactured exports (measured on a per capita basis) from the United States between 1980 to 1991.

The argument in this chapter is that, with supporting government policies, because Japan has made major strides in developing manufacturing technology and advancing S&T infrastructure, its industries have reached a level where quality-based market expansion far outweighs unfair trade practices. Japan's position in its historical alliance with the United States, however, has meant that it be on the receiving end of political pressure. Although some external criticism is healthy and can be conducive to constructive change, Japan has been subjected to much more pressure for change than can be justified by the available evidence.

Notes to chapter 4

1. Pat Choate, *Agents of Influence* (New York: Knopf, 1990).

2. One Congressperson (Bentley) commented a few years ago that: 'As the world can plainly see now, the web of deception that is [sic] Japan runs throughout the business culture--from manufacturing *keiretsu* all the way up to banks, securities firms, politicians, and the omnipotent Finance Ministry.' See *Business Week*, September 23, 1991, p. 8.

3. *Economic Report of the President* (Washington, D.C.: U.S. Government Printing Office [USGPO], 1991), p. 112.

4. See Clyde Prestowitz, *Trading Places* (New York: Basic Books, 1988), pp. 127-128; and Giovanni Dosi, Laura D'Andrea Tyson, and John Zysman, 'Trade, Technologies, and Development: A Framework for Discussing Japan,' in *Politics and Productivity*, eds. Chalmers Johnson, Laura D'Andrea Tyson, and John Zysman (New York: Harper Business/Ballinger, 1989), pp. 11-13.

5. Phyllis Genther, 'Japanese Government-Business Relations,' in *Japan's Economic Challenge*, (Washington, D.C.: USGPO, 1990), p. 48.

6. A methodologically weak attempt to persuade policy makers and other interested persons that government policies cannot promote growth in high technology areas as effectively as the private sector in Japan or elsewhere can be found in Scott Callon, *Divided Sun: MITI and the Breakdown of Japanese High-Tech Industrial Policy, 1975-1993* (Stanford, CA: Stanford University Press, 1995).

7. Martin Kenney and Richard Florida, 'How Japanese Industry is Rebuilding the Rust Belt,' *Technology Review*, February/March 1991, p. 25.

8. See Anthony DiFilippo, *From Industry to Arms: The Political Economy of High Technology* (Westport, CT: Greenwood Press 1990).

9. Japan Economic Institute of America (JEI), *Yearbook of U.S.-Japan Economic Relations in 1981* (Washington, D.C.: Japan Economic Institute of America, 1982), p. 46; and JEI, *Yearbook of U.S.-Japan Economic Relations in 1982* (Washington, D.C.: Japan Economic Institute of America, 1983), p. 46.

10. Calculated from U.S. Department of Commerce, *United States Trade Performance in 1988* (Washington, D.C.: USGPO, 1989), pp. 84, 87.

11. Nagayo Homma, 'The American Revisionists' Hostile View of Japan,' *Economic Eye*, Winter 1989, pp. 19-21.

12. For a brief but useful discussion of the Ministry of International Trade and Industry's (MITI) influence on Japan's economic development

see Toshinori Honma, 'Administrative Department Store: Has the Driving Force Behind Japan's Post-War Success Lost Some of Its Clout?' *By the Way*, May/June 1994 (vol. 4., no. 3), pp. 31-38.

13. Hugh Patrick, 'Japanese High Technology Industrial Policy in Comparative Context,' in *Japan's High Technology Industries*, ed. Hugh Patrick (Seattle, WA: University of Washington Press, 1986), pp. 9-10, 23.

14. See Daniel Okimoto, *Between MITI and the Market* (Stanford, CA: Stanford University Press, 1989), pp. 78-81. This is different than the *hojokin* financing of the past when companies had to make repayments on successful projects.

15. See Anthony DiFilippo, 'Why Is Japan Winning the High-Tech Battle?' *Forum for Applied Research and Public Policy*, Fall 1992, pp. 15-20; and U.S. Congress, Office of Technology Assessment, *Commercializing High Temperature Superconductivity* (Washington, D.C.: USGPO, 1988).

16. See 'U.S. Still Most Trustworthy, But Slipping,' *The Daily Yomiuri*, May 3, 1994, p. 3.

17. George Friedman and Meredith LeBard, *The Coming War With Japan*, (New York: St. Martin's, 1991), pp. 6-7, 381.

18. See, for example, Okimoto, *Between MITI and the Market*, pp. 64-65.

19. A Statement on National Policy by the Research and Policy Committee of the Committee for Economic Development, and a Statement by the *Keizai Doyukai* (Japan Committee for Economic Development), *Japan in the Free World Economy* (New York: Committee for Economic Development, 1963).

20. Organization for Economic Co-operation and Development (OECD), *OECD Economic Surveys 1991-1992, Japan* (Paris: OECD, 1992), p. 90. Although protecting American markets from the claimed 'dumping' practices of Japanese businesses is not something the public hears about too much today, since policy makers and other observers focus mainly on the alleged closedness of Japan's domestic markets, there is still government concern, as evidenced in a recent U.S. International Trade Commission (ITC) decision. See 'ITC Rules Printing Press Dumped,' *Washington Post*, August 22, 1996; and Jiji Press, 'ITC Rules Japanese Printers Dumped in the U.S.,' *The Weekly News*, August 22, 1996 accessed on the World Wide Web.

21. *Japan Economic Almanac 1993* (Tokyo: *Nihon Keizai Shimbun*, Inc. 1993), pp. 26-27.

22. See Subcommittee on Unfair Trade Policies and Measures, *1996 Report on the WTO Consistency of Trade Policies by Major Trading*

Partners (Tokyo: Ministry of International Trade and Industry, 1996) , pp. 114-115; also see Subcommittee on Unfair Trade Policies and Measures, *1994 Report on Unfair Trade Policies by Major Trading Partners* (Tokyo: World Trade Organization Committee of the Industrial Structure Council, 1994), p. 156.

23. Subcommittee on Unfair Trade Policies and Measures, *The 1995 Report on WTO Consistency of Trade Policies by Major Trading Partners* (Tokyo: World Trade Organization Committee of the Industrial Structure Council, 1995), pp. 50-51, 221-225.

24. Subcommittee on Unfair Trade Policies and Measures, *Report on Unfair Trade Policies by Major Trading Partners* (Tokyo: Uruguay Round Committee of the Industrial Structure Council: May 11, 1993), pp. 125-126.

25. Naohiro Amaya, 'The Causes and Cures of Trade Friction,' *Economic Eye*, December 1980, pp. 24-28.

26. See JEI, *U.S.-Japan Economic Relations Yearbook, 1984-1985*, (Washington, D.C.: Japan Economic Institute of America, 1985), pp. 30, 6-78; Laura D'Andrea Tyson, *Who's Bashing Whom? Trade Conflict in High-Technology Industries* (Washington, D.C.: Institute for International Economics, 1992), pp. 58-66; and *Japan Economic Almanac 1990* (Tokyo: *Nihon Keizai Shimbun*, 1990), pp. 36-37.

27. OECD, *OECD Economic Surveys 1991-1992, Japan*, p. 86.

28. See Robert Lawrence, 'Imports in Japan: Closed Markets or Minds?' *Brookings Papers on Economic Activity*, 2:1987, pp. 517-548; Edward Lincoln, *Japan's Unequal Trade* (Washington, D.C.: Brookings Institution, 1990); and Bela Balassa and Marcus Noland, *Japan in the World Economy* (Washington, D.C.: Institute for International Economics, 1988), pp. 49-76.

29. See Gary Saxonhouse, 'The Micro- and Macroeconomics of Foreign Sales to Japan,' in *Trade Policy in the 1980s*, ed. William Cline (Washington, D.C.: Institute for International Economics, 1983), pp. 259-304. Also see Robert Baldwin 'Commentary,' in *An American Trade Strategy*, eds. Robert Lawrence and Charles Schultze (Washington, D.C.: Brookings Institution, 1990), p. 198.

30. OECD, OECD *Economic Surveys 1991-1992, Japan*, p. 87.

31. Fumihiro Goto, *Is the Japanese Market Really Closed? A Critical Review of the Economic Studies*, (Tokyo: Research Institute of International Trade and Industry, December 1991).

32. See Ryutaro Komiya and Kazutomo Irie, 'The U.S.-Japan Trade Problem: An Economic Analysis from a Japanese Viewpoint,' in *Japan's Economic Structure: Should It Change*? ed. Kozo Yamamura (Seattle, WA:

Society for Japanese Studies, 1990), pp. 87-92.

33. Goto, pp. 16-17. The U.S. econometric model belongs to Robert Lawrence.

34. See 'Policy Statement on the Openness of Japanese Foreign Investment Policy,' in *Measures for Promoting Foreign Direct Investment in Japan* (Tokyo: MITI, January 1992).

35. From a statement by a Sales Representative for the American Ergonomics Corporation, Sausalito, California, July 12, 1993 (MITI Documents).

36. From 'Are We Abandoning Our Computer Industry to Japan?' (MITI Documents, n.d.).

37. See Lee Iacocca, 'Taking Care of Business: The Japanese Must Open Their Markets,' in *Vital Speeches of the Day* (Mount Pleasant, South Carolina: City News Publishing Company, March 1, 1992), pp. 295-299.

38. Carla Rapoport, 'You Can Make Money in Japan,' *Fortune*, February 12, 1990, pp. 85-92.

39. See National Research Council (NRC), *Learning the R&D System: Industrial R&D in Japan and the United States* (Washington, D.C.: NRC, 1990), p. 28.

40. See Global Competitiveness Corporation and Technology International, Inc., a report prepared for the National Science Foundation (NSF), *Survey of Direct U.S. Private Capital Investment in Research and Development Facilities in Japan*, NSF 91-312 (Washington, D.C.: NSF, 1991).

41. Francois van Remoortere and Donald Ketley, 'Starting a R&D Laboratory in Japan,' *Research Technology Management*, November-December 1990, pp. 37-40.

42. The Swiss firm that conducted the analysis was BERI. See Komiya and Irie, 'The U.S.-Japan Trade Problem,' pp. 83-84.

43. See Subcommittee on Unfair Trade Policies and Measures, *Report on Unfair Trade Policies by Major Trading Partners* (Tokyo: Uruguay Round Committee of the Industrial Structure Council, June 8, 1992); also see Japan's unfair trade reports for 1993 and 1994.

44. National Science Board (NSB), *Science and Engineering Indicators - 1993* (Washington, D.C.: USGPO, 1993), pp. 447-448.

45. This data is from the U.S. Department of Commerce, accessed on the World Wide Web.

46. Robert Lawrence, 'Japan's Different Trade Regime: An Analysis with Particular Reference to *Keiretsu*,' *Journal of Economic Perspectives*,

Summer 1993, p. 7.

47. William Zeile, 'Merchandise Trade of U.S. Affiliates of Foreign Companies,' *Survey of Current Business*, October 1993, pp. 59-60.

48. Ibid., p. 57.

49. Data from JETRO (Japan External Trade Organization) and accessed through the World Wide Web.

50. Kenneth Flamm, 'Making New Rules: High-Tech Trade Friction and the Semiconductor Industry,' *The Brookings Review*, Spring 1991, pp. 22-29; and 'Chip Pact Set by U.S. and Japan,' *The New York Times*, June 4, 1991, pp. D1, D7.

51. U.S. Congress, Office of Technology Assessment, *Contributions of DOE Weapons Labs and NIST to Semiconductor Technology* (Washington, D.C.: USGPO, September 1993); and The Economist Intelligence Unit, *Japan Country Report*, no. 1 (London: The Economist, 1991), p. 26.

52. See 'Foreign Chips Share in Japan Tops 20% in Last Qtr. of 1993,' *The Daily Yomiuri*, March 20, 1994, p. 5.

53. *Japan Economic Almanac 1994* (Tokyo: *Nihon Keizai Shimbun*, 1994), p. 17.

54. See Anthony DiFilippo, *'Handōtai Kyōtei: Encyō no Hitsuyou Nashi'* ('The Semiconductor Agreement: No Need for Extension'), *Nihon Keizai Shimbun*, June 22, 1996, p. 28. The semiconductor agreement between Japan and the United States reached in the summer of 1996 contains no numerical targets and places the majority of the monitoring work on the shoulders of individuals in the private sector.

55. See 'Prodding Japan, Still,' *The Washington Post*, June 30, 1996, p. C06; and 'U.S. Business Leaders Still Dissatisfied with Degree of Market Access,' *The Japan Times*, July 9, 1996, p. 2.

56. See Dick Nanto, 'Japan's Industrial Groups, the *Keiretsu*,' in *Japan's Economic Challenge*, pp. 72-88; and Michael Gerlach, '*Keiretsu* Organization in the Japanese Economy: Analysis and Trade Implications' in *Politics and Productivity*, pp. 141-174.

57. Tsuneta Yano Memorial Society, ed. *Nippon: A Chartered Survey of Japan, 1989/90* (Tokyo: Kokusei-Sha Corporation, 1989) p. 55.

58. OECD, *OECD Economic Surveys 1991-1992, Japan*, p. 84. Another business arrangement that exists in Japan is the *dango*, which could be conceived as a *keiretsu* in miniature. The *dango* produces a complementary effect in business relations, with contract assignment carried out in a somewhat informal, though still binding, manner. See Prestowitz, *Trading Places*, pp. 179-180; and *Japan Economic Almanac 1990*, p. 37. Like the

keiretsu, the *dango* is seen by some Americans as an obstacle to the penetration of the Japanese market by foreigners.

59. Nanto, 'Japan's Industrial Groups, the *Keiretsu*' in *Japan's Economic Challenge*, pp. 86-87; and Gerlach, '*Keiretsu* Organization in the Japanese Economy,' in *Politics and Productivity*, p. 171.

60. See *The Japan Letter*, no. 416, January 15, 1991.

61. Sung Yoon Cho and Constance Johnson, 'Japan's Foreign Investment Laws and the T. Boone Pickens Case,' in *Japan's Economic Challenge*, pp. 154-173.

62. Anthony DiFilippo, 'Reindustrialization Politics: U.S. Industrial Policy at a Comparative Disadvantage,' *Research in Politics and Society*, vol. 3., *Deindustrialization and the Restructuring of American Industry*, eds. Michael Wallace and Joyce Rothschild (Greenwich, CT: JAI Press, 1988), pp. 269-289.

63. Amaya, 'The Causes and Cures of Trade Friction', pp. 24-28.

64. See 'Japanese Offering Ford a Loan,' *The New York Times*, July 28, 1994, pp. D1, D7; and 'Japan Hardens Stand on Trade Talks,' *The New York Times*, September 14, 1994, p. D20.

65. OECD, *OECD Economic Surveys 1991-1992, Japan*, p. 88.

66. See NSF, *Survey of Direct U.S. Private Capital Investment in Research and Development Facilities in Japan*, pp. 2-7, 69-71. This NSF-commissioned study concludes (p. 72) that: 'U.S. multinational corporations *do not appear* to have met intolerable or even difficult restrictions in making R&D investments in Japan, and the huge majority who have followed this path indicate that their investments have been worthwhile and merit expansion.' (Emphasis added.)

67. Concerning this debate see Lawrence, 'Japan's Different Trade Regime,' pp. 3-19; and Gary Saxonhouse, 'What Does Japanese Trade Structure Tell Us About Japanese Trade Policy?' *Journal of Economic Perspectives*, Summer 1993, pp. 21-43.

68. During the course of these talks the U.S. delegation identified the *keiretsu* system as an obstacle to trade, maintaining that well-established corporate connections 'promote preferential group trade, negatively affect foreign direct investment in Japan, and give rise to anti-competitive business practices.' Quoted in Nanto, 'Japan's Industrial Groups, the *Keiretsu*,' p. 73.

69. See Susan Helper, 'How Much Has Really Changed Between U.S. Auto Makers and Their Suppliers?' *Sloan Management Review*, Summer 1991, pp. 15-28.

70. Data from GATT (General Agreement on Tariffs and Trade), accessed

on the World Wide Web.

71. According to the Japanese government, targeted, unilateral measures are being used by the United States on Japan. Only some textile products now still encounter voluntary export restraints with the United States. See Subcommittee on Unfair Trade Policies and Measures, *1994 Report on Unfair Trade Policies by Major Trading Partners*, p. 24.

72. Statements in this paragraph based on data calculated from NSB, *Science and Engineering Indicators - 1993*, pp. 160-163, 440-441.

5 Science and technology problems in the bilateral relationship

The end of the Cold War has significantly altered relations between the world's two remaining superpowers, Japan and the United States. Whether accepted or not, the end of the Cold War marked the conclusion of Japanese paternalism. While many Japanese could breathe a sigh of relief that the Soviet threat which they had long considered a fundamental danger was no longer present, many of them also understood that the removal of this perceived military problem exposed the relations between their nation and the United States to much greater scrutiny. For the United States, there has been an acceleration in what began as a slow effort to redefine national security, with much more attention focused on the economic problems that confront it. In doing this, the United States has had to engage Japan with increasing determination. On the one hand, this determination centers on developing and maintaining a new containment policy designed to deflate Japan's economic dynamism. On the other hand, this determination relates to the acquisition of Japan's science and technology (S&T).

It should be stressed that these changes did not emerge because of the dissolution of the Soviet Union. Rather, this historic event hastened these changes, since they had been in the making for some time. While interdependency is the optimal condition in the relations between the United States and Japan, this too easily dismisses superpower discord between capitalist nations and, importantly, the ongoing efforts to establish leadership which depends increasingly on the possession, management, use, and exchange of S&T. This chapter mainly analyzes the ways in which S&T and, in particular, the bilateral problems that it has engendered have figured prominently in the recent development of U.S.-Japan relations.

Suddenly, there was Japan

The heightened interest of the United States with the economic aspect of national security emerged slowly during the Reagan years and became somewhat more evident in the Bush administration. Reflecting this economic concern for national security is the expanded responsibilities of the Commerce Department's (renamed) National Institute of Standards and Technology, the establishment of its Advanced Technology Program, the appointment of an undersecretary of technology also within the Commerce Department, and in the White House itself, the official development of a structured national S&T policy.[1] Many have even considered industry-led consortia, e.g. SEMATECH, to be part of this new wave. These and others S&T initiatives directly connect to policy planners intentions to spur economic growth by increasing the competitive spirit of industry.

More than its predecessor, the Clinton administration has highlighted the linkage of economic security to national security. The Clinton administration has also endorsed an expanded science-based technology policy specifically to address problems facing the economy and industry. To a considerable extent, the subdued S&T policy initiatives of the Reagan-Bush administration and the relatively more aggressive approach advocated by the Clinton administration have been reactions to Japan, America's chief industrial rival. Although Japan has been carving out a growing industrial niche for itself for some time, serious and sustained policy reaction by the United States only appeared beginning in the 1980s.

The development of Japanese S&T policy to the level it has reached today has a strong strain of inevitability attached to it. As we have seen in Chapter 3, Japan has long sought a powerful technology base, to which it has been determinedly working to strengthen with science. Moreover, though not sufficiently defined, global leadership has become an objective of a Japan that has established itself as a economic superpower. That Japan has worked to expand and better the outcomes of its increasingly integrated and very advanced S&T network, while noticeably increasing its foreign aid, indicates that it is very aware of its position as an economic superpower. Most importantly, this empirical reality, combined with the disappearance of the Soviet Union, enhances Japan's potential to pursue a leadership position in the global political economy.[2]

We will save the detailed discussions of whether or not Japan is likely to assume this global responsibility for subsequent chapters. For now, let us emphasize that despite the general realization of the 'Japanese miracle,' U.S. policy makers and many other observers have mixed perceptions of Japan's aggregated strengths in industry and S&T. On the one hand, there is some realization that Japan's accomplishments are genuine. The view on the other hand, however, is the prevailing one, and it attributes some sense of 'wrongness' to Japan's accomplishments. Indeed, since the early 1980s the American position on Japan has been increasingly decrying its business and political undertakings by pointing to its tendency to engage in perfidious behavior -- hardly a way to maintain solid, problem-free relations. Not only have Japan's products been 'dumped' in the United States, but its domestic markets are not open enough. More recently, complaints have surfaced concerning the inability of the United States to access some of Japan's advanced products and research.

Sending 'quality' signals

Japan's industrial and S&T strengths should not be minimized as they typically are in the United States. To capture markets and advance S&T, Japan has done many of the things expected of a country in its position.[3] It is important to emphasize too that along with Japan's continuing concern for expanding and improving its S&T base, there is indisputable evidence of its increased commitment to perform more advanced basic research. Interest in improving both the quantity and quality of basic and advanced research is evident in the work being supported by universities, national research centers, industry, and government in Japan. These efforts are paramount if Japan is going to expand its leadership in the global political economy.

Already Japan has been inching toward global leadership in environmental matters. A global monitoring project within Japan's National Institute for Environmental Studies has as its objective the construction of a completely automatic 'greenhouse gas monitoring station' that when completed will be the first of its kind in the world. Another of the institute's seminal projects will connect 'ozone laser radar' to a satellite in the mid 1990s.[4] While the involvement of industry in basic research projects extends far beyond environmental projects, companies are well connected to these national and global concerns. For example, the electronics giant Matsushita is working with the Japan Environment Agency on the ozone sensor, known as the Improved Limb Atmospheric Spectrometer, which will be attached to a satellite.

Over and above the fact that proportional to its economy, Japan's research and development (R&D) effort is greater than that of any other industrial country in the world, it has in recent years been transporting results from its domestic S&T efforts to the United States. Patents granted to the Japanese *in the United States* are now considerably higher than those awarded to inventors from other nations; by the end of the 1980s almost half of foreign patents granted in the United States were to Japanese inventors. Currently receiving almost 22 percent of *all* patents granted in America, the Japanese share relative to that for the United States grew from about 6 percent in 1970 to 41 percent by 1991. Since the late 1970s Japanese patent growth in the United States has occurred in sixteen industrial areas; significant increases have appeared in both general manufacturing and in more advanced technology areas, including office, computing, and accounting machines, professional and scientific instruments, and motors vehicles.[5]

Considering that direct manufacturing investments by Japan in America have only begun to be appreciable in recent years, it is quite significant to note that by 1991 the Japanese were the second largest (behind the British) foreign owners of high technology companies in the United States. As of 1993 the Japanese owned more computer hardware and telecommunications companies in the United States than businesses from any other foreign nation, and they were strong in several other areas, such as electronic components, and biotechnology. Moreover, R&D investments in the United States by the Japanese have risen dramatically in recent years, increasing in real (1987 dollars) from $231 million in 1984 to $1,072 million by 1990.[6]

For sure, the Japanese are not inclined to move the most advanced domestic S&T work out of their country. So Japanese firms do not perform cutting edge research in the United States. The Japanese have nonetheless been much more involved in research and related activity in the United States to support their manufacturing or transplant investments. The very recent perceived 'move' on the U.S. economy by Japanese companies, however, does not signify a collective attempt to undermine American manufacturing.

On the contrary, to use the case of Japan's automobile transplants, Japanese auto executives were initially disinclined to produce in the United

States. The main reason for this was because of the large capital investment that this would entail and the potential adverse affect it would have on profits. Transplants appeared as a way to deflect complaints relating to excessive Japanese exports to the United States, and thereby prevented the increased use of American protectionism.[7]

Concerned with expanding markets, Japanese firms have therefore invested in the United States as a politically practical way to meet their expansion objective. In doing this, they have markedly increased their American patent activity. Japanese firms have also increasingly relied on research work in the United States, as well as the results of S&T activity developed in Japan.

It is important to stress that in the second half of the 1980s many Japanese firms began devoting an equal or greater amount of company funds to R&D than to capital expenditures.[8] Thus many Japanese companies have metamorphosed themselves by becoming knowledge-based establishments. This type of corporate transition has permitted Japanese companies to become more competitive in today's global economy and has also established the technical groundwork necessary to meet the challenges of the twenty-first century.

The concern with constantly improving the entire S&T base and the success in doing this is reflected in the dramatic growth of Japan's technology know-how (e.g., licenses and patents) imported by the United States in recent years. Even though the value of U.S. exports of technology know-how to Japan still exceeds that leaving Japan and coming to America, since 1981 the trend has been to import more from Japan. In 1981 Japan exported $187 million (real 1987 value) worth of technology know-how to the United States; this increased to $606 million by 1990.[9]

Bilateral politics in transition

While Japan's early export activities in the United States were in technology-driven industries such as textiles, machine tools, steel, automobiles, and televisions, it is important to stress that as time passed its products became more and more dependent on technology. These products, and others, entered the U.S. market in increasingly large numbers. In the final years of the Carter administration the main concern of U.S. policy makers with regard to Japan was how to deal with its growing exports to America. Within the political context of the perceived threats associated with the Cold War, this economic problem seemed little more than an annoyance that could be wiped away by internal policy correctives and by diplomacy. Certainly intra-alliance rivalries based on knowledge rather than arms was not even considered a likely outcome of the Cold War.

By 1980 the United States was involved in what could be called industrial self-assessment, at both the executive and legislative levels of government. Confronted with the hard reality that American economic policies were largely ineffective, policy makers considered the deleterious repercussions of industry's lost competitiveness (subsequently overlapping with the deindustrialization theme). At this time a rather extensive, yet short-lived, discussion emerged concerning the creation of a government-supported industrial policy. While this debate was largely in the context of the generic implementation of industrial policy, nonetheless the House Ways and Means Committee on Trade issued a report on the first day of October 1980 which specifically explored the benefits of learning from the policy initiatives adopted by Japan to improve the competitiveness of the U.S. computer and semiconductors industries. This report followed on the heels of another House Ways and Means Committee report whose bipartisan members concluded that with some exceptions in agricultural and high technology products Japanese markets are 'reasonably open.'[10] Although they were very much bothered by the trade imbalance, in other words, some policy makers realized that structural problems existed in the United States and that they were contributing to the industrial malaise.

When the Reagan administration assumed control of government, things did not change much toward Japan -- at least not at first. The Reagan administration initially did not have a structured Japanese policy and therefore had no good alternative but to follow the initiatives of the Carter White House which was to contend with the import problem, particularly automobiles. Begun during the Carter presidency, the initial report of a committee analyzing the economic relations between Japan and the United States known as the Wisemen's Group, and made up of both Americans and Japanese, was published in January 1981. For the most part, the new Reagan administration endorsed the report, which among other things emphasized the need to establish a more cooperative bilateral relationship and stressed that even under the most ideal conditions of access to the Japanese market, the U.S. trade deficit with Japan would still not be eliminated since American economic problems (as well as other basic disparities between the two nations) precluded this from taking place. However, after a second Wisemen's Report appeared in September 1981, the group's *raison d'etre* had disappeared because the Reagan administration had made the access of American exports to the Japanese domestic market the keystone of bilateral trade relations.[11] This decision was a crucial turning point in the bilateral relationship since the policy thrust shifted more decisively toward managed trade rather than dealing seriously with the structural problems in the American economy.

The problem of access to Japan's domestic markets had been broached and reviewed critically by some policy makers before this time, but by the end of the first year of the Reagan administration's tenure it had become a major policy instrument used by the United States. Meanwhile, America proceeded with a military build-up that produced another of the Cold War's high points. The combination of demanding market access while avoiding government-supported industrial policies to improve the nation's competitiveness aggravated the fault line in U.S.-Japan relations. Japan was continually being asked to give in to American demands, while the United States avoided making structural economic repairs that involved government planning. Although the resulting fissures widened as the trade problem with Japan worsened, the global disturbances aroused by the intensification of the Cold War became the political patchwork that covered, albeit superficially, the intra-alliance discord.

The evident discord, nonetheless conveyed two important messages, one coming from the United States and the other from Japan. The United States wanted to maintain the lead position in all facets of the alliance relationship. In regard to trade, this meant that the United States sought to increase its penetration of the Japanese market, not by legitimately improving national competitiveness, but by demanding greater access. This was especially problematic because it was precisely at this time that Japan was increasingly demonstrating the capability to build and sell high quality technology-based products. The Japanese had spent the past three and a half decades learning how to build better and better products, all the while incorporating as much advanced technology as they could into what they were doing. Their determination to experience continued success meant that over time compliance with the United States would become less likely.

However, at the time this did not mean that Japan would blatantly defy U.S. demands. Japan initially responded to the market access pressures placed on their nation by pointing out that U.S. policy makers needed to scrutinize more completely what had been happening to Japanese businesses in their country since they were not always unimpeded in America's domestic markets.[12] Still, because the alliance had deep roots and since

Japan continued to perceive the Soviet Union as a threat, it made concessions. Japan began to remove domestic trade barriers in the early 1980s in compliance with the Tokyo Round agreement.

While Japanese officials were at work trying to appease U.S. policy makers, a political cauldron had been created because America's economic policy became increasingly focused on the type of access desired in Japan's domestic market. Thus U.S. pressures went from general discussions on access to Japanese markets to specific product demands that would increase American exports to Japan. This was reflected in the MOSS (market-oriented sector selective) discussions. Begun in 1985, the main purpose of these discussions was to redirect U.S. market-opening pressure on Japan, especially in high technology product areas. Three of the four areas included in the MOSS discussions -- telecommunications equipment, pharmaceuticals and medical implements, and electronic goods -- were in the advanced technology grouping; only forestry products were not within this category. By creating a policy structure for *identified* high technology products that American businesses could export in greater amounts to Japan, U.S. policy makers had moved beyond prior demands of general access.

Unrelenting, targeted U.S. pressures on Japan for increased access to its domestic market surfaced again the following year in the semiconductor trade agreement, an accord that literally managed bilateral trade for a decade in a very critical high technology product area. Then during the Bush administration, the Structural Impediments Initiative (SII) evolved, squeezing the trigger a little harder on the political gun being held to Japanese policy makers' heads. Conceived by U.S. policy makers in 1989, SII was somewhat unlike both MOSS and the semiconductor trade agreement since these negotiations earmarked nontariff trade barriers rather than specific commodities. By January 1991 the United States added to its list of proposed changes to the government of Japan the elimination of the *keiretsu* and the increased effectiveness of its Anti-Monopoly Law.[13]

Through all of this, confusion had to be a part of the national mindset in Japan because market access only became a major policy matter at the beginning of the Reagan administration, and this after the import problem was deemed an incomplete way to compensate for America's trade problems with Japan.[14] Similarly, attacks on Japan's industrial policy, and on the existence of the *keiretsu* had by the end of the decade equally confounded the Japanese, since for sure neither of the two were recent creations. Lambasting Japan for its aggressive export policy was one thing, but attacking what the Japanese understood to be a legitimate business culture during a time when success was plainly evident served to reaffirm earlier observations made by the Japanese that America's own problems were the motivation behind its complaints and demands for change. As the Japanese saw it, by the end of the 1980s virtually anything that could appear to strengthen America's argument against their nation could and would be used.

Japan moved through the 1980s with an evolving resolve not to comply fully with the economic demands of the United States -- just what one would expect from an emerging superpower. The experience and accumulated confidence that came with continued economic and industrial success and an appreciation of the nation's improving technological and scientific competence were enough for most Japanese policy makers to rationalize their toughening position toward the United States. Even though Japanese policy makers recognized the legitimacy of some U.S. complaints, and understood that they had indeed overprotected their domestic market in the past, for the most part, they viewed America's demands on their nation as unquestionably excessive and its criticisms as exaggerated.

Another angle

As the Reagan and the Bush administrations continued to hammer away at the market access and import trade issues, what became more and more clear was that Japan was able to manage and develop technology, including the technology know-how that it continued to purchase from the United States. For the United States, this corresponded with two very important issues involved in U.S.-Japan relations. The first was increased U.S. efforts to access from Japan very advanced and necessary products, and the second was to acquire its S&T.

During the second half of the 1980s it had become evident that Japanese expertise in some advanced technology products had reached a level of development that surpassed that which existed in the United States. This meant that U.S. companies had to rely on foreign firms to supply them with advanced items. Semiconductors, semiconductor equipment, and some computer items have been particularly in demand and therefore important to U.S. companies.

As a result of a request by then Senator Bentsen, the Government Accounting Office (GAO) conducted a survey which included questioning knowledgeable persons from 59 American corporations in industries related to semiconductors, two government laboratories that purchase semiconductor products, 6 Japanese firms that produce semiconductor items, and officials from Japan's Ministry of International Trade and Industry (MITI).[15]

Some American companies had complained that Japanese firms were not granting them timely access to semiconductors and related material, and also to some computer equipment. These complaints specifically alleged that Japanese producers did not sell to U.S. companies, preferring to distribute the most advanced products to Japanese firms first. The fact that these complaints directly contravened earlier American concerns relating to excessive Japanese exports to the United States could therefore be overlooked on this matter. Since these critical technology-based products were not available in the United States, without them American companies faced severe limitations in both domestic and foreign markets. Thus this exclusionary argument helped to make the 'unfairness theme' even more elaborate: that Japan's efforts to export to the United States were excessive was supposedly only true in some areas like automobiles and steel; when it came to advanced semiconductors and related items Japan's export machine shut itself down.

As for the results of the GAO survey, they were, in a word, inconclusive. Of the U.S. firms surveyed, 85 percent of them were buying from Japanese companies. Approximately 50 percent of these companies reported to the GAO that they had not experienced any problem, or that they could not provide an example of one. Understandably, both the representatives from the Japanese companies interviewed by the GAO and officials at MITI indicated that the American claims were without basis. In any case, the survey's results do not provide sufficient evidence that Japanese producers had completely shut out American corporations on a regular basis.

An important corollary to the U.S. objective of acquiring advanced Japanese products should be considered here. Gaining swift access to them is fine for U.S. firms. But acquiring the science-based technology that led to the development of these products is fundamentally more important. There is, in other words, the definite desire on the part of the United States to obtain Japan's 'generic technology' or 'precompetitive research.' Thus obtaining Japan's S&T, whether product embedded or not, is another facet of the market access issue.

Boosting morale, a last resort?

In the face of the relative decline of America's S&T edge -- reflected in both its efforts to obtain Japan's advanced research and the erosion of its market position in some areas -- there appeared a continuing effort to raise America's expectations. For example, a former Reagan administration chairman of the president's Council of Economic Advisers has argued that because many U.S. companies had imposed needed self discipline during the 1980s this would gradually be alchemized into the basic business elements essential for 'filling in' the hollowed corporation. Corporate America was preparing to do this since it had been lowering manufacturing costs, devoting more attention to quality, and, significantly, pushing up real research and development (R&D) spending, actually to the point where it was higher than government expenditures for the first time in four decades. Together these three forces would reestablish the presence of U.S. corporations in the international global arena, while forcing American consumers to reconsider domestically built products.[16]

Perhaps these three important interacting forces would have worked had there been much less competition from Japan. Supporting this competition has been an ongoing business practice in Japanese corporations known as *kaizen* (constant improvement). The swift productivity growth for which Japanese corporations are well known relies in part on *kaizen*, which in the company environment gives some responsibility to every committed individual, from top management to workers.[17] Thus product quality improvements and attention to costs are manufactured by company-wide *kaizen*.

The fact that U.S. firms began increasing real expenditures for R&D during the 1980s and had indeed outspent the federal government in this is a useful statistical harbinger for corporate America only if one ignores the *kaizen* practice, which includes the strong commitment to research by Japanese companies. Not doing this, however, drastically changes things. While Japanese corporations spent less than U.S. firms during the 1980s, as we showed in Chapter 3, measured in real (1985) dollars their R&D expenditures were increasing faster than American companies.

To whatever extent then U.S. corporations implemented corrective measures during the 1980s, at the very least, these were offset by what had become the standard business behavior of Japanese companies. What had become virtual constants in corporate Japan -- attention to product quality, improvement in work activities and company coordination, efforts to keep costs down, and certainly, the continuous expansion of the S&T base -- had been relegated to a lesser degree of importance by too many American companies.

Accessing research

The notion of 'symmetrical access' that literally evolved out of the broader market access issue has set the stage for more serious problems in Japan-U.S. relations. In the second half of the 1980s U.S. interest in the quality of Japanese research had reached the point where the desire to obtain relatively easy access to it was more important than it had been just a few years earlier when the concern revolved mainly around technologies for military use (see below). While on the one hand, many Americans have charged that Japan, given its economic power, is doing an insufficient amount of basic research, on the other hand, there have been ongoing efforts to access its advanced work.

In 1987 members of America's National Academy of Science and the National Academy of Engineering met with representatives from the Japan Society for the Promotion of Science to discuss problems relating to the reciprocal acquisition of each nation's research work. This was the first substantive signal that the arms race was beginning to give way to the race for S&T, for at this meeting the term symmetrical access first appeared.[18] Later concerns about Japan and America's S&T systems focused on how they could become mutually beneficial to the bilateral relationship; however, such an outcome would occur only by taking the right policy steps.[19]

What are the right policy initiatives from the U.S. side, however, are not necessarily in line with the thinking in Japan. This in no way diminishes the point that in recent years American and Japanese concerns with S&T have not only become very important but they will become increasingly so in a post-Cold War environment. This is precisely because of the realization that S&T is a critical part of the economic fabric of an advanced industrial economy, since it contributes to the control of markets, and for the United States, because it also has military significance.

Critical of Japan for exploiting its basic research and mindful of its weakened hold on high technology markets, the United States has sought access to Japanese research with more enthusiasm since changes were made at the end of the 1980s in the Japan-U.S. Science and Technology

Cooperation Agreement. In the process of doing this, the United States has proposed the acquisition of Japan's generic technology which has the potential for industrial applications. This means, in other words, that there emerged the desire to access the research of private-sector corporations.[20]

As a result, this has added a new dimension to the existing problems in U.S.-Japan relations. Protectionism, managed trade, and even the continuing possibility of a heated trade war aside, symmetrical access is a very sensitive issue in superpower relations because it involves property rights and since it connects directly to markets in an increasingly competitive global environment. Whichever nation prevails here depends more and more on which one is better able to manage and protect S&T, thereby reducing market competition from the other.

Thus the relatively swift growth of Japanese company-financed research in recent years which to some extent has contributed to increased market shares has meant that more caution must be exercised before access to this work is permitted. That Japanese corporations have long financed nearly all of the nation's industrial research is no longer the central point, however. The problem has gone well beyond this in recent years.

The increased emphasis on basic research that has taken place in Japan has not been restricted to government initiatives. Institutions of higher education perform the largest proportion of basic research in Japan. However, through the 1980s and into the 1990s Japanese companies have noticeably increased basic research work. (See Figure 5.1.)

Part of the quicker growth evident in company-financed R&D in Japan relative to that in the United States had been the result of increasing spending on basic research. Real (1982 dollars) basic research performed by Japanese companies more than tripled between 1980 and 1990. Significantly, while real basic research performed by companies in Japan increased by 32 percent between 1987 and 1990, in the United States this work carried out by corporations remained unchanged. Between 1990 and 1993 the performance of basic research performed by U.S. companies declined in real (1987) value by 12 percent.[21]

That Japanese companies cut overall R&D spending for the first time in the early 1990s because of the severe recession only tells part of the story. Increasing the expectation of the corporate community that R&D spending will experience a rebound is that, despite the recession, the number of researchers employed by private companies continued to grow; significantly, by 1994 their numbers had increased by over 26,500 from 1992. The Science and Technology Agency (STA), moreover, has stressed

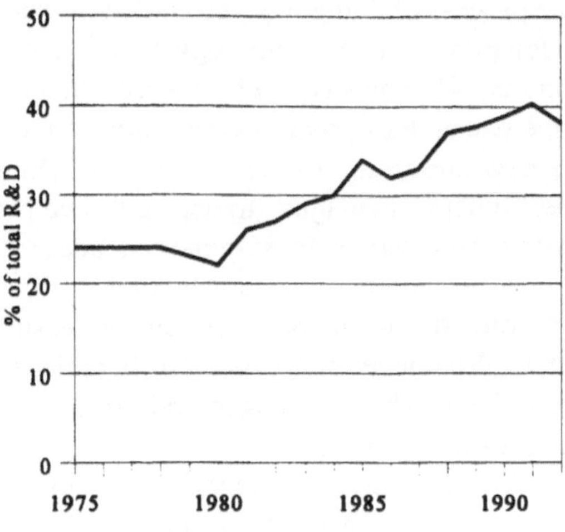

Figure 5.1 Basic research in Japan, company performed, 1975-1992

Sources: Calculated from STA, White Paper on Science and Technology, 1994. Note: data for the years 1976-1979 is estimated.

that both the government and the private sector need to strengthen basic research. This should not be interpreted to mean that investment in basic research was altogether avoided because of the recession in Japan. Indeed, in real (1990) yen basic research spending on natural science projects in Japan increased steadily each year from 1991 to 1993, growing by more than 10 percent during this period.[22]

In addition to the concern with basic research, little uncertainty should exist that the quality of this work has improved over time. Current arguments concerning the relatively low quality of science in Japan are inconsistent with U.S. attempts to pry precompetitive work out of company research laboratories there. The growth and improving quality of basic research performed by Japanese companies is a particularly troublesome matter for the United States because of its difficulty to access compared to government or university work. Private companies in capitalist societies have property rights, which they are likely to safeguard with vigilance in an increasingly competitive global environment.

Concerned with maintaining the property rights of private companies, as we noted in Chapter 3, the Ministry of International Trade and Industry (MITI) recently established its 'techno-globalism' program.[23] This and similar initiatives allay some of the bilateral friction between Japan and the

United States concerning access to advanced industry research. One of the early projects to come out of MITI's techno-globalism is the International Cooperative Research Program for Intelligent Manufacturing Systems (IMS). Because the purpose of the techno-globalism program is to share Japan's S&T achievements with other nations, IMS, a project focusing on the development of new and very complex production technologies,[24] helps to deflect criticism directed at the difficulty of acquiring access to its advanced industrial research.

Nonetheless, MITI draws a visible line between Japanese research carried out behind company walls and that performed outside of them.[25] MITI's support of Japanese firms is especially problematic for the United States which continues to pound hard on the symmetrical access drum -- and for good reason. Japanese corporations have not only become very concerned with basic research but their share of the nation's total work in this area has become noticeably bigger since the mid 1970s. By 1990 Japanese companies were performing nearly two fifths of all basic research in Japan, up considerably from 24 percent in 1975.[26] (See Figure 5.2.) This is very much unlike U.S. firms which perform only about 20 percent of the nation's total basic research. Thus U.S. pressures on Japan to permit access to its advanced research to a large extent suggests the intent to secure the property rights of Japanese corporations.

Today's research- and technology-based problems are markedly different than in the past when Japan imported large amounts of technology know-how from American companies. Then U.S. corporations permitted this to increase profits. Moreover, U.S. firms were under no pressure from Japan to export technology know-how, were not exporting precompetitive research to an advanced industrial nation, and had no idea whatsoever that they were placing in jeopardy their market position.

Playing the military card

Japan's steadily improving S&T infrastructure became importantly connected to the military objectives of the United States. More than a decade ago American policy makers tied access to Japanese technology to burden sharing. On this matter burden sharing went beyond America's conventionally understood 'free ride' diatribe, which at its core stressed that the protection of Japan was almost entirely the responsibility of the United States and that, accordingly, the Japanese needed to increase defense

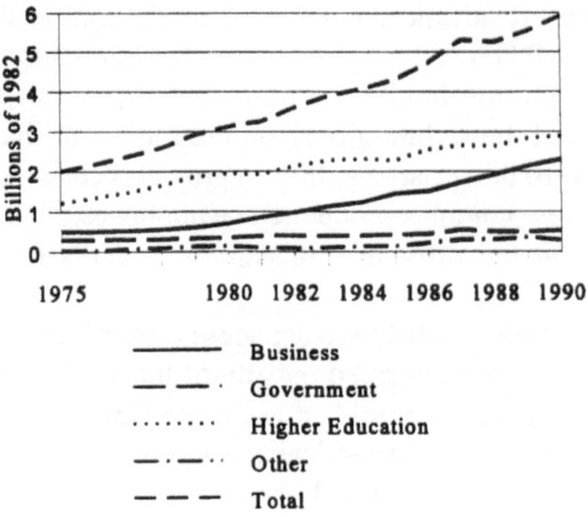

Figure 5.2 Japanese basic research by performance sector, 1975-1990

Sources: Ibid; NSF data; OECD, Basic Science and Technology Statistics; and OECD's PPP rates. Note: data for the years 1976-1979 is estimated.

spending. This (technology) form of burden sharing also went beyond America's growing insistence that the presence of U.S. troops in Japan should be a financial responsibility borne by the Japanese. Rather, this form of burden sharing emerged because U.S. policy makers came to understand that to win the arms race, which was becoming increasingly dependent on the technology race, access to Japan's technology was necessary.

The initial signs of America's intention to introduce technology burden sharing appeared in 1980 at a bilateral discussion bearing the name Systems and Technology Forum. By 1981 American policy makers indicated to Japan that it should no longer expect to get a 'free ride' from the United States. Then Secretary of Defense Weinberger told a high ranking Japanese military official that Japan could be of great assistance by exporting to the United States advanced technologies in the areas of semiconductors, precision equipment, and communications.[27] American policy makers also suggested that access to Japanese electronics and laser technologies, among others, could be useful in supporting America's military capability. While part of the Japanese government felt comfortable with the U.S. suggestions, this was by no means true of all policy makers; nor was it widely accepted by an uncertain public.[28]

Although the Mutual Defense Assistance Agreement of 1954 permitted the sharing of military-related technology and information, the Japanese government later eliminated this possibility. MITI took the initial step in this direction in 1967 when it adopted the Three Principles on Arms Exports. This policy prohibited military exports to communists nations, countries facing sanction by the United Nations, and any nation either engaged in or suggesting the potential to be engaged in war making activities. In 1976 the Japanese government broadened the three principles policy to include all nations, which also restricted access by the United States.

When the United States asked for exemption from the policy in the early 1980s, there was not a rush to meet its request. Within the Japanese government the most steadfast political groups opposed to changing the policy were the Japan Socialist Party and the Japan Communist Party. Japanese agencies with responsibilities connected to this matter were not in complete concurrence. The Ministry of Foreign Affairs favored giving an exemption to the United States, as did some of the staff of the Japan Defense Agency. However, others in the Defense Agency opposed the exemption, preferring to use domestic talents and technical expertise to advance Japan's security capabilities. MITI, most notably, objected to the exemption. MITI officials saw military-related technology transfers to the United States as undermining the proprietary rights of Japanese companies who had made significant progress over the years in commercial areas.[29]

The Nakasone cabinet drove harder for the U.S. exemption than its predecessor, the Suzuki government. In fact, Suzuki himself objected to the 'free ride' label given to Japan by the United States. Less than two months after becoming prime minister in late November 1982, Nakasone and his cabinet sanctioned military technology transfers to the United States, but only 'in principle.' This meant that transfers of military technologies to the United States were not rushing out of Japan. Prime minister Nakasone's decision to bypass the three principles policy was met by much political opposition within the Diet. In an attempt to pacify the opposition and to reach a compromise with it, the Nakasone government provided the assurance that weapons exports would still be prohibited and that only military technologies would be considered for exportation to the United States. American policy makers preferred the least cumbersome process which would allow expeditious military and dual-use technology transfers. As it turned out, a five-member Joint Military Technology Commission comprised of three Japanese (an official from MITI, one from the Ministry of Foreign Affairs, and one from the Japan Defense Agency) and two individuals representing the United States (one from the American embassy in Japan and one from the Mutual Defense Assistance Agency in Japan) became responsible for making the decisions of whether or not to permit dual-use technology transfers to the United States.[30]

Despite this apparent success of the Reagan administration, there was the realization that actual military technology transfers would not occur immediately; however, U.S. policy makers did anticipate the more timely transfer of 'dual-use technologies.' While American policy makers worked to hasten these transfers, the Japanese were dragging their feet. Dual-use technology transfers were slower than anticipated by the United States for two reasons. First, Japanese companies were somewhat disinclined to provide technologies with commercial applications, for they feared that this would come back to haunt them in the markets. Secondly, there was some reluctance in Japan to promote an image among its neighbors that it was countenancing military aggression. Meanwhile, for the United States a major problem emerged: as policy makers continued to press Japan for more dual-use technologies, uneasiness was increasing in America that it was becoming too dependent on Japan's technologies for the operation of its military systems.[31]

It is in this context that the recent technology-based problem centering on transfers emerged in U.S.-Japan relations. Because technologies and the results of advanced research are more likely to 'spin on' to the commercial sector today, this problem continues to unsettle relations between Japan and the United States. This problem then has a direct connection to bilateral trade disputes, to global market control, to the recent symmetrical access debate, and obviously to military matters. Perhaps most importantly is that as we approach the twenty-first century S&T-based matters are very likely to become even more disturbing to the relations between these two superpowers.

Already we are seeing the effects of science-based bilateral friction. The growing attention being given to basic research in Japan has aggravated U.S.-Japan relations over and above the more conspicuous trade dispute problem. Indeed, the controversies centering on 'reciprocal access,' and 'symmetrical access,' in the recent dialogue between the United States and Japan indicate the seriousness of matters relating to the availability and protection of S&T. As Japanese companies push forward with basic research, the question of access will continue to be very important.

Compounding the seriousness of this problem is that in the United States most basic research is performed in universities and is therefore considerably more accessible than that done in Japan by private companies which, as we have noted, have been doing a proportionally increasing share of this work. Companies are more likely to perform basic research if it offers the potential to be commercialized, whereas this is less true for universities. Thus Japanese firms feel that exposing their most advanced work to foreign competitors jeopardizes their market positions in the global economy. The result of all of this is to throw into doubt the superiority of American S&T, for if such were the case the continuing pressure by the United States on Japan to access its technology and its advanced research would not be evident.

Dual-use technologies and bilateralism

Too often when we hear the term dual-use our thoughts center on a company, laboratory, or other institution that develops a technology that can serve both military and civilian needs. Americans do not like to think of their S&T as less sophisticated than any other nation; this, most assuredly, applies to that used in military systems. However, as we have just seen, since the early 1980s dual-use has meant more than just technological work performed in the United States. Indeed, the quality of Japan's technology made it a potential dual-use provider since the early 1980s when there emerged the recognition that civilian technologies in a number of areas had moved ahead of those developed in the military sector.

Suggesting the strength of Japan's technology are two comparative assessments of U.S. and Japanese technologies conducted by MITI for the years 1983 and 1988 and covering sixteen important industrial areas, all with unequivocal military applications. Included in these assessments are semiconductors, communications satellites, advanced composite materials, fine ceramics, and photovoltaics. After a review of the second assessment's findings, MITI determined that Japan had made very significant advancements in technology development, clearly rivaling that of the United States.[32]

S&T assessments performed in the United States by the Department of Defense (DOD) similarly indicate very appreciable progress by the Japanese in a number of key dual-use areas. As early as 1984 the DOD identified 15 dual-use technology areas where Japan had either taken the lead or was demonstrating high level technical competence to the point of equality with the United States.[33] More recently, the DOD has stated that 'Japan has a significant lead in digital superconducting devices and systems,' and that the Japanese 'have developed niobium [an element used in metals and even superconductivity] technology to a high level of sophistication, and have overcome the long-term stability problems of the lower melting point metals which accounted in part for the abandonment of this technology by U.S. industry.'[34]

We have already seen in Chapter 2 that the DOD's post-Cold War objective for S&T is 'technological superiority.'[35] The key point reflected in this objective is that based on the performance of American S&T development, the DOD cannot efficiently manage a dual-use agenda with U.S. industry alone. Thus the underlying question is not whether the DOD requires access to Japan's S&T, but rather *how much* access is necessary to maintain technology superiority.

Figure 5.3 graphically conceptualizes our argument. That Japanese technologies are necessary to make U.S. military systems work is evident from a number of different sources. Former Secretary of Defense, Richard Cheney while in office remarked that: 'The United States is already dependent on Japan for numerous critical weapon systems components that would be essential in the prosecution of any prolonged conflict. There is great concern that this dependence could, in the future, extend to other technologies as well.'[36] Although not a long conflict for sure, during the Persian Gulf War Japanese industries provided critical technologies necessary for U.S. military systems; at the Japanese Embassy in Washington, D.C. an official there spoke of Japan as a 'very reliable source' of technology for U.S. military systems.[37] A leading Japanese politician goes much further than this. He maintains that his nation's semiconductor producers supply the technical core to many of America's military systems, since they manufacture much of the advanced microelectronic components that provide the sophisticated capabilities to U.S. weapons.[38]

The importance of semiconductors to the U.S. military objective of technological superiority and the dependence of America's defense systems on Japanese industry needs further analysis. Contrary to the popular rationalization which maintained that SEMATECH (the semiconductor consortium) was begun to improve the competitiveness of the beleaguered U.S. semiconductor industry which, in turn, would support national security (defense) objectives, there was another major reason for its appearance. SEMATECH's creation was also the result of the growing concern within the DOD, which through its Defense Advanced Research Projects Agency has financially assisted this association, that the foreign dependence of key

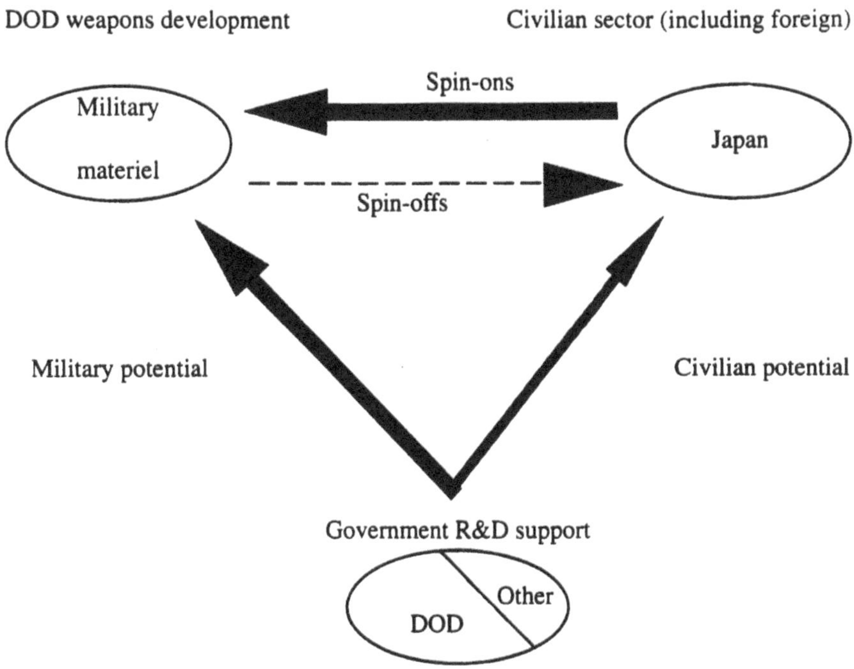

Figure 5.3 Dual-use technologies

components necessary for military systems was rapidly becoming a serious problem.

The Defense Science Board recommended in its report on semiconductor dependency in early 1987 that a semiconductor manufacturing institute be created[39]-- hence SEMATECH. While the extent of foreign (Japanese) semiconductor dependency is unknown even to DOD officials, there is complete agreement that a significant amount of it exists. Moreover, company members of SEMATECH have emphatically stated that the DOD benefits the most from the program.[40]

In what areas does this dependency exist? Certainly in the area of ceramic packages for semiconductor chips there has been considerable dependence, and memory capacity for supercomputers has relied entirely on Japanese sourcing.[41] The DOD acknowledges that 'Japan is believed to be ahead of U.S. efforts in GaAs [gallium arsenide] materials and circuit fabrication' and 'dominates in memory device manufacture, and in most aspects of microelectronics manufacturing.'[42] In 1991 the GAO pointed out that Japanese firms controlled 90 percent of the American market for semiconductor packaging devices, and that key functional semiconductor parts, including packaging housings, of the M1 Abrams tank came from foreign sources.[43] Japanese firms are the sole manufacturers of an advanced etching process (electron cyclotron resonance, ECR) used in the production of semiconductors. They control the world market for the silicon used in the semiconductor wafer, and manufacture a higher quality product than their German counterparts. In addition to semiconductor-related merchandise, U.S. dependence on Japanese technologies is also very evident in active matrix liquid crystal display screens, a product area which Japanese producers nearly dominate the world market. The Japanese are also leading producers of computer disk and laser printer technologies.[44]

Perhaps most interesting is that not only is the extent of dependency not known for semiconductors, as well as for other industrial areas not already mentioned such as advanced machine tools and robotics, but the government's 'buy American' policy generally loses track of what occurs in the procurement process beyond the prime contractor. Thus there is a rather large amount of uncertainty regarding the extent of foreign sourcing. On the one hand, policy makers and DOD officials are concerned with foreign, and particularly Japanese sourcing; on the other hand, there is no reliable way to determine the degree of foreign dependence.

The DOD directly procures only about ten percent of the military's equipment needs; the rest is done by contractors.[45] In addition to the shortcomings of the 'buy American' guidelines, prime military contractors have not had detailed policies with regard to foreign sourcing. According to the GAO, General Dynamics Land Systems Division does not have any policy to track foreign sourcing. While McDonnell Douglas does employ a guideline that fifty percent of some material come from domestic sources, this policy has not been well scrutinized. The company, for example, has not monitored this guideline for its many subcontractors. Both General Dynamics and McDonnell Douglas acknowledge that they do not keep an inventory of critical parts, which means they cannot do much about reducing foreign sourcing.[46]

Technology dependency and technology access come together in the FSX fighter dispute. This disagreement dates back to 1987 when the United States demanded the FSX project (created to provide Japan with a new fighter) be a co-production effort, rather than simply a Japanese one. The dispute escalated when the United States became alarmed that co-production would provide Japanese access to its aerospace technologies which, some critics stressed, could very well lead to the erosion of still another American high technology industry. This led the United States to insist that there be limits placed on how much advanced technology would be made available to Japan.[47]

In 1995 the United States and Japan were arguing over the ownership of and access to several advanced technologies. The DOD maintained that the United States should have access to eight technologies in the possession of Japan because they were actually American technologies which had been further developed by the Japanese. Japan, on the other hand, maintained that they were exclusively Japanese technologies.[48] In any case, the fact that this dispute involved a number of advanced technologies that at the very least had been taken to higher levels of development by the Japanese became a very serious bilateral issue central to the 'national security' of both nations.

Secretary of Defense Perry recently made it evident that he was not particularly pleased with the flow of civilian technology from Japan. The intention of his 'technology-for-technology' plan has been to encourage the bilateral exchange of technology: civilian technology from Japan to the United States, and military technology from the United States to Japan. Even though the officials in Japan's Defense Agency have recently endorsed a bilateral exchange of several important technologies, the problem remains that Japanese companies are not in any way obligated to share advanced commercial technology with their U.S. competitors.[49]

U.S. pressures, Japanese reservations

Our discussion comes together not in a causal manner but certainly in a way that we can appreciate the thrust of U.S. policy toward Japan and the development of S&T there. Pressure to gain access to Japan's dual-use technologies occurred initially because of the U.S. recognition that high quality work useful to America's military systems was being performed there. Determined to win the arms race which was increasingly being nurtured by advanced technology, U.S. policy makers saw Japan as an ally indebted to America. The Japanese, however, only saw part of this picture, notwithstanding the polite, support-America appearance commonly projected by many policy makers. To many Japanese there were limits to technology sharing. This has been true particularly when it involves private firms and their prototype, science-based technologies. Moreover, even today there remain strong reservations in Japan about supplying dual-use technologies to the United States, since it gives the appearance of violating the nation's prohibition found in its constitution against promoting war. Some Japanese policy makers rationalize this by pointing out that the technologies are developed for commercial use and it is the United States that incorporates them into military systems.[50] Unlike other foreign nations such as Great Britain and France where the sourcing of technology can be military based, acquisitions from Japan are mainly commercial, which defense contractors then 'spin on' to military systems.

For the most part, therefore, Japan has had both political and economic reasons for not completely welcoming ongoing U.S. requests for dual-use technology. Commercial technologies that are incorporated into military systems outside of Japan rub against the grain of public opinion and create problems within the political arena as well. Since Japanese technologies are largely commercial by design, the identifiable prospect of market erosion also creates strong reservations when it comes to transferring them to an industrial rival.

The U.S. desire to acquire Japan's dual-use technologies is a definitive indicator of past policy failure. Because spin-offs have not been sufficient to keep America's commercial engines roaring at full speed, and since civilian advancements have moved ahead of military ones in many areas for at least a decade, the only practical option for policy makers in the United States, short of demilitarizing and redirecting government funds, has been to turn to Japan for help. In the recent past U.S. industry simply was not able to provide all of the vital technology necessary to keep the United States ahead in the arms race; and in the aftermath of the Cold War, it is not singularly prepared to maintain technological superiority for the military.

Earlier in this chapter we discussed the complaints of several U.S. businesses which charged that Japan was not permitting American companies timely access to some advanced technologies. We showed then that there is not convincing evidence that this practice has been as widespread as alleged by some American companies. But this does not suggest that expeditious and unimpeded access has always been granted to American companies. Just as the evidence is not sufficient to establish that Japanese companies have repeatedly denied U.S. firms access to advanced technologies, neither is it clear that they are supplying them everything when requested. For sure, Japanese firms, like companies elsewhere, are not going to 'give away the store' to their competitors, though this does not mean that they would be unwilling to do business with them.

For the Japanese, moreover, this is an issue which goes beyond merely divulging company secrets and market competition. Japanese companies have a long history of working within the *keiretsu* (or industry groups). There is a discernible fidelity to member companies within the several operating *keiretsu* in Japan -- perfectly natural and fully acceptable within its business milieu. Although the American interpretation of this is 'unfairness,' since it seemingly focuses on the protection of products created by the fear of reduced market share while patently disregarding the economic principle of buyer access based on availability, for the Japanese another concern is paramount. This is the cultural allegiance to the *keiretsu*, which satisfies the needs of member companies, and also its workers. (In Japan workers have benefited from the lifetime employment that has been an integral part of the *keiretsu*, which also uses personnel transfers in the expanding business network as a reward to loyal, competent and long-time employees.) While the *keiretsu* are not exclusionary organizations, that is, in regard to foreign businesses as we saw in Chapter 4, member allegiance can include supplying industrial materials or products on a priority basis to buyers within the organization.[51]

To illustrate the connection between the protection of advanced technology and the *keiretsu* we will take a closer look at a small part of the bilateral problem in semiconductors. One of the most frequently mentioned semiconductor items allegedly denied American companies are steppers, which are used in bringing the circuit design to the disk or wafer. Steppers are not abundantly produced in the United States, but they are in Japan where semiconductor-related companies belong to vertically structured industry groups. Moreover, Japanese manufacturers have a reputation for building steppers that are the highest quality in the world. The United States has a very high foreign dependency rate for stepping aligners.[52]

Steppers have commercial as well as military applications. By providing America unimpeded access to the most advanced steppers, while permissible because of changes in Japanese policy concerning dual-use technology transfers to the United States, Japan's control of an important advanced technology market could easily be compromised. At the same time such access would undermine the loyalty system inherent to the *keiretsu*. In addition to both of these reasons, expeditious access may even contribute to the existing public outcry opposed to exporting technologies that have direct military applicability.

Assessing the fissure

Japan has clearly industrialized to the point where the results of its S&T are in demand by the United States. In some cases this demand is based on the fact that there is simply no other practical choice, since producers other than the Japanese either do not manufacture the items, or when this does occur, they are of lower quality. As we have seen in this chapter, this demand has recently moved beyond product access and has been elevated to requests for precompetitive or generic research, which may include basic research. The main point here is that Japan's industrial S&T development has evolved to where its real or potential application can have use in either commercial or military work. In both forms, there are great strains being placed on Japan-U.S. relations.

With regard to military-related technologies, only in the past few years have the Japanese become somewhat more accommodative to U.S. requests for dual-use technologies. More recently, an important debate in Japan has been whether or not there is a need to change the nation's peace constitution to permit direct participation in international military activities.[53] Although U.S. criticism of Japan on the burden sharing issue has existed for many years, this debate has recently been refueled, thus becoming more visible because of American complaints of Japan's belated 'dollar diplomacy' response to the Persian Gulf War.

There is currently some political support in Japan for constitutional change. But this is offset by political opposition both inside and outside of the Diet. There is also considerable public antipathy toward altering the peace constitution, as we will see in detail in Chapter 7.

Reflecting some of the resistance in Japan to create an 'exterior' military posture and the reluctance to resurrect the specter of militarism and alarm it Asian neighbors is that the initial effort to pass the Peacekeeping Operations Bill (a measure seeking to sanction the sending of Japan's Self Defense Forces abroad for international security purposes) was not successful. Committed to creating the legality of an international security presence by Japan and to increase the nation's chances of obtaining a permanent seat on the United Nations Security Council, the Liberal Democratic Party led by former prime minister Miyazawa, and with the support of the Komeito Party and the Democratic Socialist Party succeeded in passing the bill in 1992. By October 1992 Japan had about 1,800 of its Self Defense Forces involved in noncombat peacekeeping operations in Cambodia. The bill includes a number of important provisions which prohibit any form of combat involvement by Japanese forces. Because of public uneasiness and vehement opposition by the Social Democratic Party of Japan and the Communist Party, the government has no immediate plans to attempt the removal of these provisions.[54]

While efforts to establish 'normal' (i.e., like Western industrial nations) peacekeeping activities in the future will definitely meet with political resistance in Japan, success in winning over public opinion to support combat-active international security theoretically will open the door even wider than it is at present for dual-use technology transfers to the United States, and could conceivably reverse the current attempt by MITI to create an Export Trade Control Order. This legislation is intended to make it more difficult than it is now to export high- and even medium-level technological products which could be used in the construction of weapons by a number of nations considered by the Japanese government to be predisposed toward military conflict.[55] For these reasons, constitutional endorsement of 'normal' international peacekeeping would be very likely to bring about additional bilateral tension, since America would not be inclined to support Japanese dual-use transfers to other nations, though it would expect expeditious access to virtually all of Japan's science-based technologies. The latter issue would dangerously interact with the prerogative of private firms to guard advanced research. Although the full extent to which all of this will aggravate superpower rivalries is unknown at this time, even under present conditions it is a certainty that the tug of war over S&T will be a continuing and major problem in the bilateral relationship.

On the commercial side, Japan's strong reservations about transferring state-of-the-art products and advanced research to the United States strikes two very sensitive nerves. The first relates to the core of any capitalist system, that is, the proprietary right of a company to restrict access to its research-related work. Permitting access even with compensation that includes a suitable profit still places a company's market share at risk. This is particularly so at a time when S&T is as critical as it is to market control in the global economy. S&T and related support to companies from MITI and other government agencies ultimately underscores the economy-wide issue which is fundamentally that easy access to advanced science-based work could place in danger the nation's standard of living.

Secondly, there are important concerns relating to the culture of business organization, the *keiretsu*. On this matter the Japanese are responding to two different obligations at the same time. On the one hand, there is the obligation to the *keiretsu* which has traditionally required member companies to support the industry group. On the other hand, there is the a lingering, though less important, obligation to the United States for past financial and continuing military support. These diametrically opposed and unequal obligations pull at the very fabric of the Japanese political economy. Thus short of the dissolution of the *keiretsu*, and assuming that blood is thicker than bilateral politics, U.S.-Japan relations will continue to be strained since many of the results of the most advanced S&T work will initially be circulated within industry groups.

Although we have typically separated the military and commercial issues related to S&T, our discussion strongly suggests that there is considerable overlap. Military and commercial S&T are not only very much interwoven, but they impact on a number of issues critical to U.S.-Japan relations such as bilateral trade, military security, economic competitiveness, and ultimately global leadership. Even though there is evidence of cooperative work on advanced technology projects between the United States and Japan at both the government and company levels, for example, an international semiconductor consortium involving IBM, Toshiba (Japan), and Siemens (Germany), and another with Fujitsu and Advanced Micro Devices (U.S.),[56] these actions fall far short of reconciling the power struggle that accompanies the bilateral relationship. This is particularly true when, as is now the case, property rights and the guarding of S&T are central to intercapitalist competition.

Notes to chapter 5

1. See Alan Bromley, Executive Office of the President, Office of Science and Technology Policy, *U.S. Technology Policy* (Washington, D.C., September 23, 1990); also see Executive Office of the President, Office of Science and Technology Policy, *Science and Technology* (Washington, D.C., March 17, 1992).

2. Related issues are discussed in Koji Taira, 'Japan, an Imminent Hegemon?' in *Annals of the American Academy of Political and Social Science*, January 1991, pp. 151-163.

3. See Anthony DiFilippo, *From Industry to Arms: The Political Economy of High Technology* (Westport, CT: Greenwood Press, 1990); and Anthony DiFilippo, 'Why Is Japan Winning the High-Tech Battle?' *Forum for Applied Research and Public Policy*, Fall 1992, pp. 15-20.

4. National Science Foundation (NSF), *Tokyo Report Memorandum*, February 9, 1993.

5. NSF, *International Science and Technology Data Update: 1991* (NSF 91-309), pp. 101-105; and National Science Board (NSB) *Science and Engineering Indicators - 1993* (NSB 93-1), p. 455. More analysis of this issue can be found in Anthony DiFilippo, 'Converting U.S. Science and Technology Resources to Civilian Use: Industrial Exigencies and Political Implementation,' in A. Brunn, L. Baehr, and H. J. Karpe, eds. *Conversion: Opportunities for Development and Environment* (Berlin: Springer-Verlag, 1992), pp. 268-281.

The growth of Japanese motor vehicle patents combined with the increased production of Nissan and Toyota in the United States provide a justification for inserting their names as substitutes for General Motors in a once hackneyed expression and thus reintroducing it as 'what is good for Nissan and Toyota is good for the economy.'

6. NSB, *Science & Engineering Indicators - 1993*, pp. 384, 470; also see Edward Graham, 'Japanese Control of R&D Activities in the United States: Is This a Cause for Concern?' in *Japan's Growing Technological Capability*, eds. Thomas Arrison, C. Fred Bergsten, Edward Graham, and Martha Caldwell Harris (Washington, D.C.: National Academy Press, 1992), pp. 189-206.

7. Concerning the Japanese automobile companies in the United States, see Robert Perrucci, *Japanese Auto Transplants in the Heartland* (New York: Aldine De Gruyter, 1994), especially pp. 2-5.

8. R&D outlays compared to capital investments for Japanese companies

during the period 1985-1987 can be found in Fumio Kodama, *Analyzing Japanese High Technologies: The Techno-Paradigm Shift* (London: Pinter 1991), pp. 18-21.

9. Technology know-how exports from Science and Technology Agency (STA), *White Paper on Science and Technology, 1992* (Tokyo: The Japan Information Center of Science and Technology, 1993), p. 260.

10. See Japan Economic Institute of America (JEI), *Yearbook of U.S.-Japan Economic Relations in 1980* (Washington, D.C.: JEI, 1981), pp. 5, 58-59.

11. JEI, *Yearbook of U.S.-Japan Economic Relations in 1981* (Washington, D.C.: JEI, 1982), pp. 6, 45-47.

12. JEI, *Yearbook in 1981*, p. 46.

13. *Japan Economic Almanac 1991* (Tokyo: *Nihon Keizai Shimbun*, 1991), p. 32.

14. See Anthony DiFilippo, 'Technology and Trade Problems in Postindustrial Economy: Japan Versus the United States,' in *Perspectives and Issues in International Political Economy,* ed. Chronis Polychroniou (Westport, CT: Praeger, 1992), pp. 150-171.

15. See Government Accounting Office (GAO), *International Trade: U.S. Business Access to Certain State-of-the-Art Technology* (Washington, D.C.: GAO, September 1991).

16. See Murray Weidenbaum, *Small Wars, Big Defense* (New York: Oxford University Press, 1992), p. 99; and Murray Weidenbaum, 'Filling In the Hollowed-Out Corporation: The Competitive Status of U.S. Manufacturing,' *Business Economics*, January 1990, pp. 18-22.

17. Masaaki Imai, *Kaizen: The Key to Japan's Competitive Success* (New York: McGraw Hill, 1986).

18. Both U.S. academies function in an advisory capacity to the federal government. See National Research Council, *Expanding Access to Precompetitive Research in the United States and Japan: Biotechnology and Optoelectronics* (Washington, D.C.: National Academy Press, 1990).

19. See Martha Caldwell Harris, 'Asymmetries and Potential Complementaries: Scientific and Technological Relations Between Japan and the United States,' in *Japan's Economic Challenge* (Washington, D.C.: United States Government Printing Office [USGPO], October 1990), pp. 275-288 (Study Papers Submitted to the Joint Economic Committee, Congress of the United States); also see Statement of Dr. Martha Caldwell Harris, Director, Office of Japan Affairs, National Research Council, *Hearings Before the Joint Economic Committee, Congress of the United*

States, Thursday December 6, 1990 (Washington, D.C.: USGPO, 1992), pp. 234-244.

20. *Japan Economic Almanac 1990* (Tokyo: *Nihon Keizai Shimbun*, Inc., 1990), pp. 93-94.

21. Calculated from STA, *White Paper on Science and Technology 1992*, pp. 248, 250 and from the Organization for Economic Co-operation and Development's purchasing power parity exchange rates; U.S. figures calculated from NSB, *Science and Engineering Indicators - 1993*, p. 334.

22. NSF, *Tokyo Report Memorandum*, December 6, 1994; and *Tokyo Report Memorandum*, December 8, 1994, p. 2.

23. See Ministry of International Trade and Industry (MITI), *Issues and Trends in Industrial/Scientific Technology: Towards Techno-Globalism* (Tokyo: MITI, June 1992).

24. *Japan Economic Almanac 1991*, p. 77.

25. MITI, *Issues and Trends*, p. 40.

26. The debate (fueled by which exchange rate conversion is the most appropriate) amongst some American economists concerning whether or not U.S. corporations are outspending Japanese firms misses the point. While everyone agrees that Japanese firms have spent heavily on R&D, the key neglected point is that they have been performing more basic research than in the past. See Graeme Browning, 'The R&D GAP,' *National Journal*, April 4, 1992, pp. 805-809 for a summary of the R&D debate between U.S. economists.

27. Isamu Miyazaki, 'Economic Prosperity or a Strong Military?' *Economic Eye: A Quarterly Digest of Views from Japan*, March 1982, p. 27.

28. See JEI, *Yearbook in 1981*, pp. 51-52; and JEI, *Yearbook of U.S.-Japan Economic Relations in 1982* (Washington, D.C.: JEI, 1983), pp. 72-73.

29. Ibid.; and James Auer, 'Defense Burdensharing and the U.S.-Japanese Alliance,' in *Japan and the United States: Troubled Partners in a Changing World*, (New York: Brassey's, 1991), p. 34 (Special Report).

30. See Report of Defense Science Board Task Force on *Industry-to-Industry International Armaments Cooperation, Phase II-Japan* (Washington, D.C.: Office of the Under Secretary of Defense for Research & Engineering, June 1984). Also see JEI, *Yearbook of U.S.-Japan Economic Relations in 1983* (Washington, D.C.: JEI, 1984), p. 63; and Michael Chinworth, with Barbara Wanner, 'Whatever Happened to Defense Technology Transfers,' in *Japan: Exploring New Paths*, ed. Jon Choy (Washington, D.C.: JEI, 1984), pp. 137-141.

31. JEI, *Yearbook of U.S.-Japan Economic Relations in 1983*, p. 63; and JEI, *U.S.-Japan Economic Relations Yearbook 1984-1985* (Washington, D.C.: JEI, 1986), pp. 58-59. At the same time that the United States was positioning itself to acquire Japanese dual-use technology, it was denying this type of technology to Japan. See JEI, *Yearbook in 1983*, p. 63; and *Japan Economic Almanac 1991*, p. 131.

32. Cecil Uyehara, 'Appraising Japanese Science and Technology,' in *Japan's Economic Challenge*, pp. 295-296.

33. Ibid., pp. 295-299.

34. Department of Defense (DOD), *Critical Technologies Plan*, May 1 1991, pp. 19-10 - 19-11; also see DOD, *DOD Key Technologies Plan*, July 1992.

35. See DOD, *Defense Science and Technology Strategy*, July 1992, p. I-2.

36. Richard Cheney, 'Competition Among Friends,' in *Powernomics*, eds. Clyde Prestowitz, Ronald Morse, and Alan Tonelson (Lanham, Maryland: Madison Books, 1991), p. 61.

37. See 'Japanese Companies, in a Nation of Peace, Are Manufacturing the Technologies of War,' *Asian Wall Street Journal*, February 11, 1991, pp. 1, 4.

38. Shintaro Ishihara, *The Japan That Can Say No* (New York: Simon & Schuster, 1991), pp. 20-22.

39. See Defense Science Board (DSB), *Report of the Defense Science Board Task Force on Defense Semiconductor Dependency*, Office of the Undersecretary of Defense for Acquisition, Washington, D.C., February 1987.

40. GAO, *Federal Research: SEMATECH's Technological Progress and Proposed R&D Program*, Washington, D.C., July 1992, p. 26.

41. DSB, *Report on Defense Semiconductor Dependency*, pp. 62, 64.

42. DOD, *Critical Technologies Plan*, pp. 2-15, 1-17.

43. GAO, *Industrial Base: Significance of DOD's Foreign Dependence*, Washington, D.C., January 1991, pp. 19-20.

44. GAO, *International Trade: U.S. Business Access to Certain Foreign State-of-the-Art Technology*, September 1991, p. 15.

45. DSB, *Final Report of the Defense Science Board 1986 Summer Study*, (Office of the Undersecretary of Defense for Acquisition, Washington, D.C., January 1987), p. 19.

46. GAO, *Industrial Base: Significance of DOD's Foreign Dependence*, p. 17.

47. *Japan Economic Almanac 1991*, p. 131. Co-production means 40 percent American and 60 percent Japanese.

48. 'A New Warplane's Murky Horizon,' *The New York Times*, January 13, 1995, pp. D1, D12.

49. 'Japan Suggests Technology Swaps with U.S. in Key Defense Areas,' *The Nikkei Weekly*, August 22, 1994, p. 4; and 'Washington Mulls U.S.-Japanese Security Changes,' *The Japan Times Weekly International Edition*, February 6-12, 1995, p. 3.

50. See 'U.S. Relied on Foreign Parts for Weapons,' *The Washington Post*, March 25, 1991, pp. A1, A17.

51. See Michael Gerlach, '*Keiretsu* Organization in the Japanese Economy: Analysis and Trade Implications,' in *Politics and Productivity*, eds. Chalmers Johnson, Laura D'Andrea Tyson, and John Zysman (New York: Ballinger, 1989), pp. 141-174; Michael Gerlach, *Alliance Capitalism* (Berkeley: University of California Press, 1992), pp. 143-146; and Dick Nanto, 'Japan's Industrial Groups, the *Keiretsu*,' in *Japan's Economic Challenge*, pp. 72-88.

52. GAO, *International Trade: U.S. Business Access to Certain State-of-the-Art Technology*, p. 8.

53. See, for example, Yamauchi Toshihiro, 'Gunning for Japan's Peace Constitution' *Japan Quarterly*, May-June 1992, pp. 159-167; also see Kenji Suzuki, 'Dangerous Syllogisms: Constitutional Revision Not a Cure-all for Japan's Woes,' *By The Way*, May/June 1993, pp. 11-15. Two other articles in this same issue (Mitsunori Takehana, 'A Matter of Interpretation: Proponents of Constitutional Reform Have Mixed Agenda with Questionable Motives,' and Susumu Nishibe, 'A Case for National Identity: Japan Searches to Define Itself Amidst Pressure for Change at Home and Abroad') also deal with the matter of constitutional reform.

54. See *Japan Economic Almanac 1993* (Tokyo: *Nihon Keizai Shimbun*, 1993); and The Economist Intelligence Unit, *Country Report, 2nd Quarter 1993, Japan* (London: the Economist Intelligence Unit, 1993), p. 12.

55. 'Recession is Longer Biggest Threat to Firms' ('MITI Watch'), *The Japan Times Weekly International Edition*, November 21-27, 1994, p. 6.

56. See *Japan Economic Almanac 1993*, p. 88.

6 Making its move: Japan as a global superpower

Policy making in Japan after the American occupation did not revolve around an internally inspired Machiavellian nucleus as the nation began to develop its industrial strengths.[1] Most especially in the past two decades as Japan steadily evolved into an economic superpower, it is incorrect to allege that industrial success occurred because of sustained and wrongful government policies in conjunction with unfair corporate practices, both of which are sometimes said to be exacerbated by national xenophobia. Not to be confused with unfairness to foreign business, it is true that with government support and cooperation, Japanese corporations have been relentlessly concerned with economic growth since World War II. The results have been near continuous economic boom and growing market shares. A concomitant objective has been to improve living conditions for the people of Japan, which with many sacrifices has been accomplished throughout the post-war period.

This chapter begins the analysis (which is completed in Chapter 7) of the interplay between Japan as an economic superpower and the changes that have taken place in the global system during the past several years. This chapter's most important message is that Japan has become well suited to be a global leader. In addition to its economic success and competency in science and technology (S&T), Japan's domestic socioeconomic accomplishments strengthen its image as a global leader. Because it is reasonable to assume that a superpower is a model for other nations, particularly developing ones in geographical proximity to it, we will first examine a few of Japan's domestic socioeconomic accomplishments. We will next turn to a source of bilateral discord not yet discussed in this work,

Japanese direct foreign investments in America, since this is another way in which Japan's economic influence relative to the United States has grown in recent years.

These economic and related developments, however, do not stand alone. The demise of world communism has not only removed the substantive foundation upon which U.S.-Japan relations were built, but it has further opened the window of opportunity which first appeared when Japanese economic power became evident. This opportunity has afforded Japan the chance to forge a new path toward becoming a *genuine global superpower* in the post-Cold War period, as this chapter will begin to argue. To have more global clout than it had during the Cold War period, Japan needs to assume more international responsibilities.

Global changes have given Japan the chance to respond independently to global conditions. No longer having to worry continuously about fitting under the American security umbrella, Japan's decisions in some important ways are being made as a world leader, not as a nation that must first fully consider U.S. reactions. While much depends on how Japan evolves as a global superpower, as we shall see, the potential is there for it to make major contributions to the structure of the world economy and political system during the remainder of this century and into the next.

A kinder form of capitalism

Along with the very probable elimination of Armageddon-like conflict, the end of the Cold War has most certainly turned the focus of attention for all nations on the prospect of economic growth. Japan has shown that a large part of a nation's population does not have to be disregarded as economic growth occurs. Western critics of Japan, however, have ignored its domestic socioeconomic achievements, as well as the real and potential effects of these accomplishments on nations struggling to meet domestic and global challenges. For several of these nations Japan is the source of a 'demonstration effect'; Japan is showing that strong economic growth *is* compatible with a relatively humane industrial society.

Unlike the United States in particular, Japan has combined economic growth with considerably more attention to the basic components necessary for individual development. Although revisionist observers emphasize the 'differences' between America and Japan and focus consistently on the Schumpetarian-oriented characteristics of Japanese capitalism,[2] none of them stress that its economic system is profoundly more equitable compared to that of the United States.

Why do the two leading capitalist nations manifest markedly different socioeconomic outcomes? Part of this answer, for sure, relates to the Schumpeterian notion of innovation. Japanese innovations (propelled by organizational and managerial efficiency, combined with technological and scientific expertise) have contributed much to the economic growth that has permitted improved living conditions. Certainly another part of the answer to this question is because of the willingness of the Japanese to use government to direct civilian-economic and S&T policy. Still another important fact is that community concern which has traditional roots in Japan is easily adapted to its cultural homogeneity. Finally, despite the dominance of the Liberal Democratic Party for decades, strong socialist opposition has been a continuing factor in Japanese politics during the post-war period. Socialist influence on Japanese politics has worked to constrain business, while attempting to foster community concern.

Japanese capitalism is more than just different from that found in the United States. It is a system where a genuine sensitivity for 'peoplism' is built into the very fabric of the political economy and is actually demonstrated in existing conditions.[3] A brief look at a few of these conditions existing in Japan is in order.

The benefits to the economy from Japan's very strong educational system have already been discussed in this work, so only a few other important points on this subject need to be emphasized here. Apart from Japan's nearly nonexistent illiteracy rate (0.2 percent), its basic educational system is recognized globally as excellent. Japan has the lowest high school dropout rate in the world (2 percent). During the 1980s over one third of Japanese students were continuing their studies beyond high school, that is, they were attending junior colleges or colleges and universities; this figure increased to over 45 percent by 1995.[4]

Significant in the context of developing global leadership skills is that in recent years Japan's educational system has been giving more attention to the promotion of peace (*heiwa*). An elementary school headmaster recently stated that: 'All over Japan, the emphasis on peace has increased.' He also points out that the Ministry of Education, Science, and Culture which provides 'The educational guidance that all teachers receive has begun to place more emphasis on peace, with more suggestions how they can cultivate students to seek peace.'[5]

In addition to a strong and increasingly responsible educational system,

Table 6.1 **Japanese income by quintiles, wage and salary income, 1976, 1980, 1991**

Average annual income in dollars by quintiles		Ratio of highest quintile to others
Year 1976		
I (lowest)	$6,368	2.7
II	8,496	2.0
III	10,156	1.7
IV	12,578	1.4
V (highest)	17,036	
Year 1980		
I	$9,845	2.6
II	13,100	2.0
III	15,855	1.6
IV	19,015	1.4
V	26,110	
Year 1991		
I	18,782	2.9
II	26,094	2.1
III	32,367	1.7
IV	39,197	1.4
V	54,515	

Sources: Calculated from Statistical Yearbook of Japan; and OECD data.

another socioeconomic accomplishment evident in Japan is relative income equity. Per capita national income now slightly exceeds that of the United States, making it the highest in the world.[6] Compared to other industrial nations, unemployment in Japan has been quite low. For the most part, the specter of job loss has not been something the majority of Japanese workers have had to worry as much about as have workers at all occupational levels in the United States. The fallout from technology attrition that has adversely affected workers in the United States and other Western nations has yet to impact on the Japanese economy, notwithstanding the fact, for example, that it has the most robotized economy in the world. Poverty is not a widespread problem in Japan. Those whose incomes are substandard are dealt with on an individual basis and are compensated by the nation's social welfare programs. In 1993, only about 1.4 percent of all Japanese households were receiving public assistance, while 12.3 percent of American families were living below the poverty line in this same year.[7]

Income is distributed more equitably in Japan than it is in the United States. Data from Japan's Economic Planning Agency show a more equitable income distribution in Japan than in the United States, as well as a continued trend of stability in income distribution since the mid 1970s.[8] Suggesting the latter is that data computed from government surveys of Japanese households show no major change over time between the highest and lowest average annual income quintiles of wage and salaried employees. (See Table 6.1.)

Compared to the United States, astronomically high executive salaries in Japan are not much of a problem. The chief executive officers of Japanese corporations make considerably (approximately four times) less than their U.S. counterparts. This therefore means that the average distance between executive pay and the average production worker in Japan is significantly less than it is the United States.[9]

It is clear from what we have briefly discussed above that Japan is truly different from the United States in a socioeconomic sense. The differences we have identified, however, are not the ones that commonly occupy the attention of Western analysts critical of Japan. Should Japan continue to maintain relative economic equity at home as its position in the international political economy becomes more and more important, then the chances improve that emerging nations, particularly in the Asian region, will look increasingly at it -- and not the United States -- as a developmental model. To the extent that other countries 'import' Japanese capitalism, including its concern for labor and its stress on national income equity, the relative global influence of the United States would decline.

More reasons for bilateral discord

Different though it is from American capitalism, the results of Japanese capitalism have become integral to the U.S. market. Despite the fact that U.S. policy makers have long been critical of Japanese business and government practices, substantial dependence on Japan is evident. This dependence ranges from many products to a number of important technologies.

Deepening this dependence is the increased economic involvement of Japanese businesses in the United States. For about two decades after World War II, Japan's foreign investments were minimal. More recently Japan's direct foreign investments have grown. In the United States they have risen sharply, more than doubling from 1987 to the present. Along with substantial investments in the banking and financial sectors of the United States, the Japanese have increased their investments in its manufacturing sector, thereby helping to fill a void left by American industry. Tables 6.2 and 6.3 show the increased interest of the Japanese in U.S. manufacturing during recent years.

The small decline in Japanese direct manufacturing investments (historical-cost basis) in the United States during 1993, as seen in Table 6.2, was the result of recession in Japan. Still, there were 1,695 Japanese transplants operating in the United States in 1993. Although this was somewhat less than in 1992, a larger number of firms in 1993 (more than half) indicated that they were profitable, according to JETRO (Japan External Trade Organization). A JETRO official stated that: 'It is unlikely that the decline in the number of factories in the U.S. will continue due to economic recovery.'[10]

Although Japan's direct investments have also increased in Europe, approximately two fifths of its total foreign investments are in the United States.[11] Why has Japanese industry turned its attention to the United States for investment? Most obviously, it is profitable. In the United States, where consumerism is extraordinarily high, investment is a practical decision.

Because of increased investments in the United States during the past several years, many more Americans now owe their livelihoods to Japanese companies. The employment of Americans in the United States by Japanese nonbank businesses grew from 427,100 in 1988 to 728,200 in

Table 6.2 Japanese direct investment in the U.S. manufacturing sector, 1987-1995

Year	Billions of dollars
1987	4.97
1988	11.06
1989	15.60
1990	17.15
1991	18.18
1992	18.78
1993	17.57
1994	20.07
1995	21.19

Source: U.S. Department of Commerce, Survey of Current Business, various issues.

Table 6.3 Japanese assets in the U.S. manufacturing sector, 1986-1992

Year	Billions of dollars
1986	11.43
1987	15.28
1988	33.02
1990	43.51
1990	65.06
1991	70.00
1992	72.04

Source: Same as in Table 6.2.

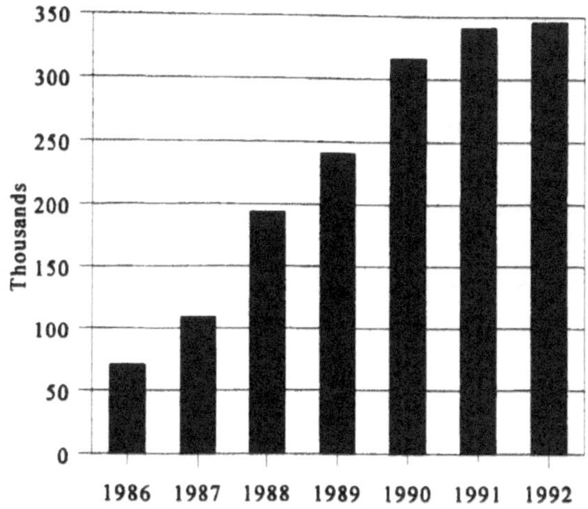

Figure 6.1 Japanese employment in the U.S. manufacturing sector, 1986-1992

Source: U.S. Department of Commerce, Survey of Current Business, various issues.

1992 to 756,500 in 1994.[12] All told, Japanese businesses are currently the second largest foreign employer of Americans in the United States, behind only Great Britain. Increased employment by Japanese businesses is evident in the U.S. manufacturing sector as well. Figure 6.1 shows the dramatic growth of Japanese manufacturing employment in the United States.

What Japanese employers lack in numbers they make up in employee compensation. Compensation to American workers by Japanese businesses operating in the United States in 1991 came to $39,287 (up from $37,470 in 1990). This was considerably more than the average for foreign nonbank enterprises which was $36,162, including the averages for British and Canadian companies, which were $32,693 and $33,517 respectively.[13]

However, the explanation for Japanese investments in the United States is not as straightforward as suggested above. Part of the answer is indeed obvious: Japan is an very advanced capitalist nation with matured industries that individually intend to expand their sphere of influence. Perceived market opportunities in the United States, whether they be in finance, services, or manufacturing, including high technology areas, will thus attract Japanese investors with capital and the requisite expertise. Japanese investors in the United States have invested heavily in plant and equipment, and have patented the necessary technologies. But this only partly explains Japanese interest in the American economy. Trade barriers to Japanese imports which stretch back to the 1970s created a 'circumvention initiative,' that is, production in the United States avoids American protectionism. The rapid appreciation in the value of the yen against the dollar that began in the latter half of the 1980s has been another critical factor accounting for the surge of Japanese investments in the United States.[14] When such an investment formula exists, as it has for Japanese investors, foreign investment will occur.

There are two direct benefits to the U.S. economy that result from Japanese investments and transplants. The first is that Americans get jobs. Here it shall not be debated whether or not the employment resulting from businesses not owned by Americans is good for the nation, but it is important that we at least briefly mention labor relations in Japanese plants located in the United States. Although Japanese investors have not fully transplanted labor relations to the United States, preferring to adapt to existing conditions, there is some evidence that worker relations are good, resulting in the desired outcome of high productivity which rivals that which exists in Japan.[15] There is additional evidence that Japanese automobile transplants in the United States have eschewed layoffs and thus far have demonstrated a commitment to job security for workers.[16]

Certainly, employee compensation helps to maintain good labor relations. As a whole, employee compensation by foreign businesses operating in the United States is considerably higher than that of nonbank American-owned companies.[17] Given the relatively higher employee compensation paid to American workers by Japanese businesses located in the United States, as indicated above, one would suspect that this true in at least parts of the manufacturing sector. It is known that the six Japanese automobile transplants in the United States, which employ approximately 30,000 workers, are paying average hourly wages which exceed regional manufacturing wages and are only two dollars below those of the unionized 'big three' American companies.[18]

The second benefit resulting from Japanese investments and transplants in the United States is that the trade data for any given industry appears better than it would without them. Much of what Japanese manufacturers produce in the United States is consumed by Americans.[19] Given the size of the U.S. trade imbalance with Japan, from one angle this is very important, since U.S. imports, at least of consumer-ready products, are lower than they would be otherwise; moreover, there is some increase in the volume of American exports.

Overshadowing these benefits, however, is the very inauspicious perspective that many Americans have of Japan. This perspective creates a substantial amount of animosity about Japanese investments in the United States, and for this reason intensifies bilateral discord. Part of this animosity relates to extant cultural differences and, because of America's past paternal relationship with what became a very wealthy Japan, another part connects to Japanese reluctance to meet U.S. expectations for sharing the military burden. Thus Japanese investments in the United States stoked the 'free-ride' diatribe, which, interestingly enough, empirical evidence does not support.[20]

Japanese investments in the United States have also created a 'virtual reality' that could arguably be contributing to the distortion of policy making in the United States. Because Japanese transplants make the trade picture for a given industry look brighter than it would without them, it becomes more difficult than it already is for interested public officials in the United States to organize meaningful cooperative policies (involving industry and labor, as well as government laboratories and universities) designed to strengthen beleaguered industries.

For example, the U.S. machine tool industry, once the dominant one in the world, has been running trade deficits for many years. Most U.S. imports of machine tools come from Japan. Recently, the U.S. machine tool industry's trade deficit had been declining. Facing a trade deficit in 1989 which stood at $1.4 billion, by 1992, after decreasing steadily, it amounted to $770 million. Protectionist measures partly account for this decline in the industry's trade deficit, but so does the existence of Japanese transplants in the United States. Japanese machine tool plants in Kentucky, California, Alabama, and North Carolina have increased the production of machine tools in the United States at the same time they have helped to improve the industry's trade imbalance.[21]

Thus government initiatives to revitalize this very important but besieged sector of the economy have been inadequate at least partly because the machine tool industry's problems appear less severe than they actually are to policy makers. This has helped those who have demonstrated a persistent reluctance to assist industry in entirely new ways remain the dominant policy making force, while those potentially more receptive to an interventionist approach are inclined to more easily accept arguments which downplay the seriousness of the industry's problems. All of this has resulted in current policy making for the machine tool industry to be not too different from that of the past.[22]

What is likely to continue to unsettle bilateral relations is the way that Japanese businesses operate in the United States. Japanese businesses in the United States are responsible for a big proportion of all American imports from Japan. A large percentage of these imports are intrafirm exchanges. In the final analysis, however, the percentages of imports that came from parent companies to Japanese businesses in America (intrafirm imports) from the late 1970s to the early 1990s were basically the same as those for German affiliates in the United States. Still, because of increased Japanese manufacturing investments in the United States in recent years, the manufactured imports of Japanese transplants operating in America have increased significantly. In 1977 Japanese affiliates operating in the United States accounted for 5 percent of the imports of all foreign-affiliated companies; by 1991 they had increased their share to 29 percent. Much more so than companies from other industrialized countries, Japanese manufacturing firms operating in the United States rely heavily on imported intermediate inputs. In 1991 the import component used in production by Japanese manufacturing companies operating in the United States stood at 28 percent (31.7 percent for all Japanese businesses), noticeably higher than the 17.3 percent recorded for all foreign-affiliated manufacturing firms (19.6 percent for all foreign industries). In two Japanese manufacturing industries operating in the United States -- computer and office equipment, and audio, video, and communications equipment -- the percentage exceeds 50 percent. In others such as machinery, both electrical and nonelectrical, and in motor vehicles, the percentages range from almost 33 percent to just under 50 percent.[23]

Thus Japanese affiliates in the United States have strong connections with businesses in Japan. Home-country supplier relations are very much in evidence for Japanese transplants. The relatively higher proportion of imported production inputs characterizing Japanese manufacturing transplants in the United States is especially ammunition for those politicians, policy analysts, and labor organizers who feel compelled to 'prove' unfairness, since they can ring the familiar jobs-lost bell to generate American animosity toward Japan.

This whole matter is further complicated by the fact that many of the Japanese industries operating in the United States are also those that rely heavily on research and development to remain competitive. Because most of the leading edge research and development is not performed in the United

States by Japanese transplants, bilateral tensions could easily worsen since cooperative projects in America would be reduced by the location of this work, or otherwise stated, the desire to restrict access to research undertaken in Japan. Along with the relatively higher import proportion used in production by Japanese firms in the United States, the performance location of advanced research can easily become another issue that could be used by those who feel threatened by Japanese companies doing business in America.

Certainly one important measure of the economic strength of industries in an advanced capitalist nation is their positions in world markets, and another is their propensity to invest in foreign countries. While Japanese industries have been developing their export capabilities for years, it is only recently where they have moved to expand significantly their foreign investment position -- a 'natural' response for industries when sufficient resources are available. What makes Japan's situation unique is that its 'new imperialism' has not relied on domestic military support. Rather, Japanese industries have used capital, technology and managerial resources to literally carve out pieces of the U.S. market. More recently, Japan's emerging global political initiatives have been partly built on these economic successes.

Japan's increased economic strength coupled with its advancing S&T capabilities make it an important force today in the global political economy. These strengths also create an enormous potential for Japan in the years ahead. Because Japan's economic power increases its political responsibilities in the world community, there is the very real prospect that under present conditions some of its global decisions will put it at odds with the United States.

The crumbling cornerstone

On May 1, 1987 former prime minister Nakasone made two very important points concerning U.S.-Japan relations during a speech given in Washington, D.C. He first emphasized that: 'The trade imbalance, which underlies our bilateral frictions, arises fundamentally from structural causes. This is precisely why the trade and economic problems between Japan and the United States are so intractable.' He then went on to state that the 'Japan-U.S. Security Treaty [is] the cornerstone of our relations.'[24] To completely understand these statements, we must first consider a few important points.

Because of the differences between the policy making processes in Japan and the United States, Japanese S&T capabilities evolved differently from those in America. This occurred despite some similarities in the consumption of technologies. A fundamental reason for the bilateral trade imbalance (but certainly not the only one) is the fact that Japan's commercial S&T capabilities have consistently improved during the post-war period, reaching a level of competence that rivals that of the United States. On the other hand, American policy makers' lack of enthusiasm for a lasting commercial S&T policy and their obsession with maintaining military superiority at any cost, helped bring about and sustain structural problems that weakened U.S. (import and export) competitiveness. Significantly, Japan's industrial S&T capabilities have been inextricably linked with its export objective: high quality exports could not have materialized without the steady improvement in its S&T capabilities.

Concerning Nakasone's second comment, while the U.S.-Japan security agreement remains intact, its political justification does not, at least not for the Japan which had previously perceived a threat from the former Soviet Union. It is true that the Japanese government still endorses the Japan-U.S. security arrangement in the post-Cold War environment; this is especially true for the Defense Agency,[25] even though today the likelihood of any major military threat directed at Japan is very remote. For its part, Japan is working to ensure that even this remote possibility does not occur. For example, in May 1993 former prime minister Miyazawa's cabinet approved a $100 million 'free aid' package to Russia in support of disarming its nuclear weapons. The formal expression of this initiative took shape in a proclamation by the chairman of Japan's Atomic Energy Commission and bears the title 'Cooperation Concerning Nuclear Weapons Disposal.'[26]

What was in the past largely a bilateral trade problem between the United States and Japan has been expanded to include issues involving access to advanced products and research, as we saw in Chapter 5. Precarious in its own right, this 'issue inflation' comes at a time when two other important things have happened to affect U.S.-Japan relations. The first is Japan as an emerging global superpower, which increases its freedom to make independent decisions. The other is the dissolution of the Soviet Union which diminishes, if not entirely eliminates, the need for the security arrangement. Thus the crumbling of the 'cornerstone' in U.S.-Japan relations makes it easier for Japan to assume a somewhat autonomous leadership position in global politics.

One important area where this could occur is in resolutions appearing in the United Nations concerned with the use of nuclear weapons. In a vote on the proscription of the use of nuclear weapons in late November 1961, Japan cast its ballot in favor of the resolution, while the United States opposed it. The gist of this and similar resolutions subsequently introduced is that nuclear weapons are a preeminent danger to human existence and therefore any nation using them is committing a crime against the peoples of the world. After the 1961 vote, including those taking place into the 1990s, Japan has opted to abstain, feeling concerned about the prospect of an inauspicious U.S. response.[27]

As a superpower in the post-Cold War era seeking to forge a unique place in global politics, however, Japan could now easily respond to a long-held national belief of working to bring about world disarmament, a conviction that stems directly from its peace constitution. This is a very serious issue, and one which we will later explore in detail, especially in Chapter 7.

Important global initiatives by Japan

What we will examine in this section are two important ways in which Japan has increased its visibility in global politics -- cooperative international S&T projects, and its willingness to assume a larger responsibility to tackle big environmental problems. We will also examine how these efforts connect to present and future U.S.-Japan relations.

Japan's commitment to expand cooperative international S&T projects -- e.g., the Human Frontier Science Program -- certainly is an important sign that it is willing to contribute to the global community. Such undertakings financed by government help to create a positive image of Japan as a responsible superpower. However, this in no way diminishes the likelihood of the continued collisions between the United States and Japan concerning access to private sector research.

Hardly anyone today would deny that the level of S&T development within a nation helps define its existing material conditions; this will become increasingly true in the future. Because Japan-U.S. relations have become more and more based on the results and impact of innovative S&T, the scramble for access to it and protection of it becomes more intense. Japan, then, is presently confronting a major dilemma: as an emerging global superpower with an exceptionally strong commercial S&T base, the expectation of it instituting international cooperative projects has become quite high. But it order to maintain the economic and industrial momentum of the past, Japan must guard the proprietary work of its corporations which are assuming a larger role in the nation's advanced research effort.

Under present conditions, the United States and Japan have no acceptable criteria for settling the big problems related to the access of advanced corporate research, and none should be expected, despite the continued existence of the Japan-U.S. Science and Technology Agreement (in effect now until 1998). The attempt to help mend what is undeniably a U.S. competitiveness problem with a dual-use policy is a telling reminder of America's failures in handling its past development of S&T. Related to this is that the desire to exploit Japan's commercial sector is inescapable for American companies desiring to strengthen competitiveness and for a government still determined to maintain high tech military systems.

Thus as Japan works to craft an image in the eyes of the world community as a major advocate of cooperative S&T with initiatives such as the International Workshops Support Program, a project which promotes the direct exchange of scientific and technological information between researchers, it has at the same time been particularly mindful of the need to safeguard the advanced research of the nation's corporations. The fact that Japan's global posture improves as it sponsors cooperative international projects is helpful, since this relieves some of the pressure it faces from the United States concerning access to corporate research work. Nonetheless, Japan-U.S. relations will be strained no matter how big of an international sponsor of S&T Japan becomes, since America will not be privy to much of the advanced private-sector industrial work.

Concerning the global environment, Japan's growing willingness to address this issue has become notable in the past few years. But before examining this matter, it would first be informative to discuss Japan's early environmental problem, as well as its emergent and increasing interest in pollution control. Japan began a sustained campaign against pollution in the 1960s, especially in the second half of the decade, while it was experiencing rapid economic growth. Indeed, it is because of this rapid economic growth, which produced toxic environmental effects, that the pollution offensive began in Japan. After establishing his cabinet on November 9, 1964, prime minister Sato quickly demonstrated a strong commitment to take on Japan's big pollution problem. With Diet support, the Japanese government passed several important and effective anti-pollution laws and implemented pollution- related initiatives. Both the Ministry of International Trade and Industry and the Ministry of Health and Welfare were especially active in Japan's campaign against pollution in the

1960s. Sparking this official interest was the growing public concern with the nation's pollution problem. Particularly responsible for attracting the public's attention was the 'big four pollution trials' which took place because of the pollution-related diseases (the *Minamata* and *Itai-itai* diseases, and *Yokkaichi* asthma). Japan's continued commitment to economic growth and concern for the environment today make it a model for industrializing nations that opt for 'sustainable development.'[28]

Not only has Japan's environmental transition been commendable, but public awareness of and concern for global environmental problems are high. A recent public opinion poll showed that 43 percent of the Japanese respondents were very concerned about international problems related to global warming produced by the 'greenhouse effect,' compared to 30 percent in the United States.[29] Currently in Japan expectations are quite high that S&T will be very important in solving global environmental problems. A survey conducted by the Science and Technology Agency showed that the leading answer of respondents was that they expected S&T '[t]o contribute to a reduction in pollution and global environmental problems.'[30]

However, Japan's current interest in confronting global environmental problems with advanced S&T has a good chance of colliding with the United States, thereby further disturbing superpower relations. Given the unimpressive American record in dealing with global environmental problems, should Japan assume the responsibility as the caretaker of these global concerns, it will have to consider the United States a transgressor. In the United States conservative business groups, along with some politicians, have resisted environmental regulation by government.

For example, as the Clinton administration prepared itself to grapple with the global warming problem, conservative forces revived their efforts to use science to capitalize on the ambiguity between the greenhouse effect and carbon dioxide waste from automobiles and industry. This retrograde effort to reshape national thinking on global warming equates inconclusive scientific evidence with the ideological position of minimal government intervention in the economy. At the same time, this effort disregards the beliefs of the majority of climatologists and rejects the unequivocal warning of the U.S. National Research Council which directs attention to the potential problems associated with global warming, as well as the advice twice given by a scientific committee working to counsel the United Nations which anticipates a rise in the earth's temperature because of greenhouse effects.[31]

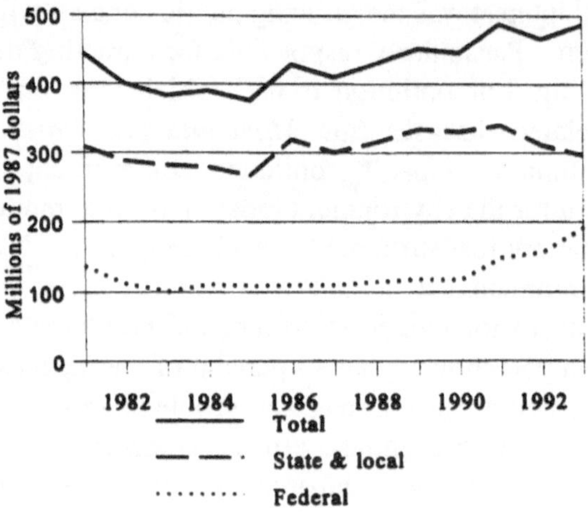

Figure 6.2 Government air pollution regulation and monitoring in the U.S., 1981-1993

Source: U.S. Department of Commerce, Survey of Current Business, various issues.

Significantly, this effort also presages a major point of political divergence between America and Japan. Recall, for example, that in Chapter 5 we briefly discussed the recent creation of the Action Program to Arrest Global Warming by the government of Japan. This is a continuing project which identifies the greenhouse effect as a very serious problem facing the nations of the world, and one which Japan is determined to mitigate in the future. Thus to the extent that Japan assumes a bigger role in environmental leadership and specifically commits its national resources to the problem of global warming, it will be confronting powerful forces in the United States determined to limit public influence in the world economy on this important issue.

Indeed, despite heightened public concern about global warming in recent years government spending in the United States to regulate and monitor air pollution has not changed in more than a decade. Figure 6.2 shows that total government spending in real (1987) dollars has remained more or less constant in the United States. Federal spending, which has been significantly less than that by state and local governments, has increased modestly in inflation-adjusted dollars since 1991.

Reducing air pollution has not been a high priority in the United States.

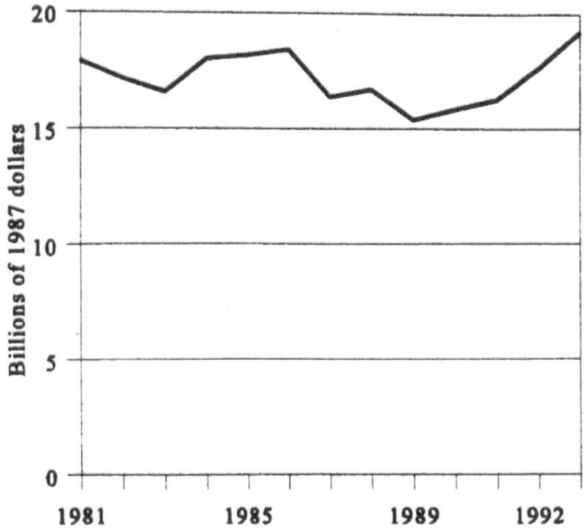

Figure 6.3 U.S. air pollution reduction, government and business, 1981-1993

Source: Same as Figure 6.2.

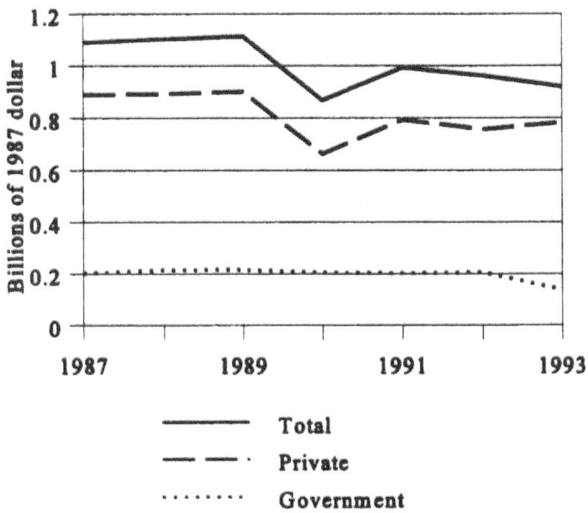

Figure 6.4 R&D spending in the U.S. on air pollution control and abatement, 1987-1993

Source: U.S. Department of Commerce, Survey of Current Business, May 1995.

There has been a negative, though rather weak, relationship between the combination of government and business spending for air pollution abatement and gross domestic product (GDP) for more than a decade.[32] Figure 6.3 shows that real spending for air pollution reduction by industry and government in the United States has stayed within a rather narrow band between 1981 and 1993. Just as significant, Figure 6.4 shows that in real dollars little change has occurred in the United States during this time in research and development spending for air pollution abatement either by government or industry.

The lack of resolve in America to confront the air pollution problem is especially important after considering that relative to other major industrial nations the United States is by far the largest emitter of harmful gases into the atmosphere, such as carbon dioxide, and nitrogen and sulfur oxides. Japan, on the other hand, emits a smaller amount of sulfur and nitrogen oxides than Germany, Canada, France, Italy, and the United Kingdom. With the exception of the United States, Japan's carbon dioxide emissions are greater than other industrial nations.[33] However, in his speech at the United Nations Conference on Environment and Development in Rio de Janeiro in June 1992, former primer minister Miyazawa reaffirmed Japan's commitment made in its Action Program to Arrest Global Warming to hold the nation's emissions of carbon dioxide by the year 2000 to near the amount discharged in 1990.[34]

Although the U.S. economy is much bigger than Japan's, the disparity between the GDPs of these two nations in no way accounts for the differences in contaminated emissions. Comparative data for 1989 show substantial differences between the United States and Japan of noxious emissions of both sulfur and nitrogen oxides into the air, even after accounting for disparities in GDP. (See Table 6.4). Concerning greenhouse gases that are generally agreed to contribute to global warming, and particularly carbon dioxide which is by far the largest of the air contaminants in both countries, emissions by the United States are substantially greater than in Japan. While it is true that chlorofluorocarbon emissions by Japan are out of line with the size of its GDP, this air contaminant makes up less than 10 percent of its total emissions of greenhouse gases. (See Table 6.5). When analyzed on a per capita basis and by per unit of GDP, Japan's emissions of carbon dioxide, nitrogen oxides, and sulfur oxides rank lower than those for United Kingdom, Germany, and the United States.[35]

The Japanese government has certainly demonstrated a heightened sense

Table 6.4 Air pollution emissions, Japan and the U.S., sulfur and nitrogen oxides, 1989

Emissions (tons)	Japan/U.S. (percent)
Sulfur oxides	*Sulfur oxides*: 4.1%
Japan: 876,000	
U.S.: 21,510,000	
Nitrogen oxides	*Nitrogen oxides*: 6.7%
Japan: 1,301,000	
U.S.: 19,290,000	
	Relative GDP: 37.7%

Source: Organization for Economic Co-operation and Development OECD Environmental Data, 1993 (Paris: OECD, 1993).

Table 6.5 Emissions of greenhouse gases, Japan and the U.S., 1990

Emissions (tons)	Japan/U.S. (percent)
Carbon dioxide	
Japan: 1,125,500,000	
U.S.: 5,066,950,000	22.2%
Methane	
Japan: 1,338,000	
U.S.: 37,000,000	3.6%
Chlorofluorocarbons	
Japan: 111,000	
U.S.: 130,000	85.4%
Totals	
Japan: 1,126,949,000	
U.S.: 5,104,080,000	22.1%
	Relative GDP: 39.3%

Source: Same as in Table 6.4.

of commitment to global environmental problems. Public expenditure in Japan for pollution abatement (all forms) as a percentage of GDP is larger than that in the United States, as well as in other major industrial nations.[36] The government of Japan has been increasing expenditures for the preservation of the global environment, allocating approximately $5.5 billion to this in 1994. Three fourths of this amount, or about $4.1 billion, was for global warming.[37]

That Japan will see itself as the international arbiter of environmental problems and especially global warming is not a remote possibility. Prior to the international meeting held in Rio de Janeiro, the secretary general of the United Nations Conference on Environment and Development requested that Japan become a leading advocate of 'sustainable development,' a growth model that literally suggests present and future concerns for the environment. Because Japan has shown that relatively low levels of contaminant emissions and the widespread use of technologies that are sensitive to the environment do not run counter to economic growth, this request is perfectly reasonable.[38]

Japan therefore is being asked to be a model to the world, including industrializing nations. Itself having experienced serious environmental problems as a result of rapid economic growth following World War II, as we have seen, Japan took important corrective steps. Largely successful in its efforts to reduce pollution, Japan has since been able to combine strong economic growth with concern for the environment. At the United Nations Conference in Rio de Janeiro former prime minister Miyazawa made it perfectly clear that Japan was ready to 'play a leading role in the international efforts for both environment and development' and maintained that it is imperative 'that environmental protection and sustainable development be achieved in tandem.'[39]

Should an international organization that seriously intends to confront global warming and other environmental problems form, which is a distinct possibility,[40] Japan's environmental interests will not be analogous to those of the United States. In this context, superpower disagreements will escalate to a higher level than they have been in the recent past. Because the United States has been skeptical of the direct connection between greenhouse gases and global warming and has resisted setting controls on harmful emissions,[41] the prospect of increased discord with Japan on the matter of global warming should not be too quickly ruled out as another significant matter unsettling the bilateral relationship.

Foreign aid and technology development

In addition to being a model for economic growth and environmental development, Japan has emerged as principal provider of foreign aid to the developing nations. Although facing ongoing pressures from the United States to increase military spending in the 1980s more than it actually did, Japan instead turned to foreign aid. Besides the domestic opposition to increasing military spending in Japan, there was also neighboring-Asian resistance. Thus increasing foreign aid provided Japan with a way to deflect U.S. criticism directed at the burden sharing matter, and at the same time its Asian neighbors felt relief in that remilitarization was not taking place.

Beginning in the mid 1950s Japan linked foreign economic aid to war reparations, distributing it mainly to its Asian neighbors. This reparation-linked foreign aid did not end until 1977. Because of its overriding concern with economic growth and expanding its export trade, Japan also connected its foreign aid to export promotion from the 1950s up to about 1970.[42]

Beginning in the late 1970s Japan's foreign policy intentions came into focus: avoid a significant military buildup and concentrate on economic assistance. Because it maintained the offensive military constraint imposed by Article 9 of the constitution, which was assiduously guarded by the nation's leading opposition party, then the Japan Socialist Party, Japan did not fully succumb to American pressures. Indeed, in 1977 former prime minister Fukuda promised that Japan would not attempt to become a major military power, but rather its global objective would be focused on economic cooperation.[43]

Cooperating with other nations by providing economic assistance is exactly what Japan did. Once a recipient of economic assistance from the United States, Japan eventually emerged as a major provider of official development assistance (ODA) in the 1980s. After taking into account the effects of exchange rates and prices, Japan's ODA increased from being under two fifths of that provided by the United States in the mid 1970s to over two thirds by the early 1980s. During the second half of the 1980s Japan's ODA approached that provided by the United States, and surpassed it in 1989.[44] Growing by almost 20 percent from 1990, Japan's ODA spending amounted to $11.034 billion in 1991, compared to $9.642 billion for the United States.[45] The socialist-led coalition government that came to power in the summer of 1994 continued to support Japan's nonmilitary financial assistance to the global community. Japan's 1995 fiscal budget called for a 7.7 percent increase in ODA; in 1995 Japan's ODA spending reached nearly $15 billion.[46]

Japan has not been intent on just providing economic aid to other nations. Rather, it has taken on the responsibility of attempting to promote both environmental security and global peace. Speaking to the Diet in April 1991, former prime minister Kaifu stated that his nation's foreign aid would be guided by a policy acceptable to Japan. Subsequent to this, Kaifu's guidelines became part of Japan's official foreign assistance policy. Approved by former prime minister Miyazawa's Cabinet on June 30, 1992, Japan's foreign assistance charter consists of a basic philosophy and four guiding principles, and is now one of the most extensive foreign aid policies in the world. The basic philosophy states that it is necessary for Japan 'to play a role commensurate with its position in the world to maintain world peace and ensure global prosperity,' and speaks to environmental and human rights concerns while endorsing sustainable development. In addition to concern for the environment, the three other principles in the charter are as follows: that ODA funds, as in the past, cannot be used for military purposes; that the trend of a recipient nation's military spending will be completely scrutinized, as well as the country's military destructive capabilities and its involvement in arms trade, with the overall objective being the encouragement of national resources for social and economic development; and lastly, that a recipient nation be actively involved in promoting democracy and a 'market-oriented economy,' while ensuring basic human rights.[47]

Japan's current ODA spending is built on the momentum established by the First Medium-Term Target, an initiative begun in 1977 that very successfully doubled foreign economic assistance by 1980 instead of in five years as originally planned.[48] By June 1988 Japan had instituted its Fourth Medium-Term Target. This target's objective was to double within five years total ODA spending over that which took place during the 1983-1987 period. A major concern of the Fourth Medium-Term target was to increase grant aid, also an objective of Japan's Fifth Medium-Term Target, scheduled to run from 1993-1997. Technical cooperation assistance, a type of grant disbursement, has been growing in recent years. A number of other kinds of aid are included within the category of technical cooperation. In addition to basic human needs in the form of medical and public health aid, Japan provides assistance for computer science work and for other sophisticated technologies. By providing technology and training to 'counterparts' in developing nations, Japan is attempting to improve general economic conditions in developing nations. Through its development technical cooperation assistance, Japan is attempting to promote industrial expansion in developing nations by nurturing the growth of research and development and by transferring technologies.[49] Besides technical cooperation assistance, Japan has provided substantial loans and other grant aid to developing nations.

Significantly, Japanese funds have also been increasing to multilateral organizations, particularly to the United Nations (U.N.). Increasing multilateral funding was a major objective of both the Fourth and Fifth Medium-Term Targets. In recent years Japan has increased its funding to multinational organizations such as the U.N.'s Environmental Program, the U.N.'s Developmental Program, the International Development Bank, and the European Bank of Recovery.[50]

The United States and Japan have already brushed elbows on the matter of influence in multilateral lending institutions. Japan's growing contributions to the World Bank increased its voting share at the expense of the United States, which did not look kindly on its reduced influence. In fact, the United States consented to Japan's additional voting shares in the mid 1980s when Japanese funding to the World Bank's International Development Agency increased but insisted that it maintain exclusive veto power in the organization. Moreover, because of its increasing financial contributions to multilateral lending institutions, there has been a good amount of pressure on Japan at home and abroad to take on a more conspicuous leadership position in them. Although thus far Japan has generally aligned itself with the United States on matters of concern to multilateral lending institutions, there certainly is no guarantee that this will not change.[51]

In any case, still other examples of Japan's multilateral assistance in recent years are its $1.2 billion contribution in 1991 to organizations that work to assist refugees throughout the world; currently, it continues to provide substantial assistance for this purpose. Japan has a major voice in the Asian Development Bank. Japan is presently the largest contributor to this development-oriented organization. Funds earmarked for the Asian Development Fund, which is directed by the Asian Development Bank, make Japan the single largest financial provider to this effort designed to assist developing nations. Japan recently instituted the Policy and Human Resource Development Fund as a functioning part of both the International Bank for Reconstruction and Development and the International Development Agency. Already financially committed to the European Bank for Reconstruction and Development, which works to help nations that have recently experienced political and economic transition, the government of Japan has also instituted the Japan Europe Cooperation Fund.[52]

Although Japan's foreign aid program is stretching beyond Asia, this region still receives the bulk of Japanese assistance. About 60 percent of Japan's ODA goes to Asian nations at the present time. However, this is down from over 90 percent distributed to this region in 1970. This decline has occurred mainly because nations in other regions such as the Middle East, Africa, and Latin America have been receiving increased aid from Japan in recent years.[53] It should also be pointed out that the purpose of Japan's foreign aid is different from that of the United States and most other industrial nations. Much of Japan's foreign assistance to recipient nations has been earmarked for economic infrastructure (approximately 50 percent), unlike the United States which assigns a relatively small proportion for this purpose (below 5 percent).[54]

Even though over three fourths of Japan's foreign assistance is in the form of untied aid, the disproportionately large share going to its neighbors definitely increases its sphere of influence in Asia. Illustrating this is the Japanese interest in ASEAN (Association of Southeast Asian Nations) -- Indonesia, Malaysia, Philippines, and Thailand -- to which Japan's ODA has helped pave the way for its foreign direct investments and for improving its trade objectives. Long major beneficiaries of aid from Japan, the ASEAN countries have recently experienced substantial growth in Japanese foreign investments. In fact, Japanese foreign investments in these nations have exceeded those of other countries. As a result of these investments, Japanese companies have been establishing production bases in these nations, thereby creating 'sourcing cites' from which Japan has increasingly been importing manufactured goods, which currently make up a significant proportion of its total manufacturing imports. Moreover, Japanese exports to the ASEAN countries have grown noticeably in recent years.[55]

Given the manifest differences in Japanese and American capitalism, this kind of 'economic vortex' created by Japan in the Asian region has the potential to become a bigger political problem than it is currently in the bilateral relationship. The increasing use of the members of ASEAN for off-shore production and exporting from this area to the United States will serve to check America's manufacturing trade deficit with Japan but could greatly exacerbate political problems between the two nations. In the post-Cold War period a financial incentive to foreign nations in some respects is a potent substitute for military security, and so the continued use of it by Japan in Asia translates into additional political leverage in this region of the world. Already we see in the newly formed APEC (Asia Pacific Economic Cooperation) forum that while Japan understands its emergent leadership responsibility in the Asian markets as being tied to regional economic cooperation, as does most of its neighbors, the United States has recently been pushing for an economic model based on the immediate and complete 'liberalization' of the area as a way to increase American exports and investment.[56] Thus American pressure to impose a 'free market' model in an area that currently endorses economic cooperation is currently another political problem facing the bilateral relationship.

Freed of past Cold War rivalries, global attention is increasingly shifting to the economic differences existing between Japan and the United States, and this necessitates changes in their political relations. Thus to the extent that Japan is able to convince other nations that its form of capitalism -- characterized by a relatively equitable income distribution, low unemployment, strong economic growth, nonmilitary foreign aid, and a rapidly expanding S&T base which has demonstrably nurtured industry -- can reap more economic benefits than the American model, current economic tensions with the United States will become very troubling political problems.[57]

Refocusing the United Nations

Bolstered by its advancing S&T competence, Japan's increasing economic influence around the globe is already being reflected in the U.N. Despite the recent innovative contributions to the U.N. that we have previously noted, Japan's influence in the U.N. is likely to grow in the next several years. Thus the real question is the extent of global leadership to be assumed by Japan through the U.N.

It is well known that Japan's financial contributions to the U.N. are second only to the United States. However, Japan's influence in the U.N. is currently growing for more reasons than this. Japan has long felt that the U.N. should provide security to the nations of the world. The end of the Cold War has created the opportunity for this to occur. Japan's emergence as an economic superpower has buoyed national confidence to the point where leadership responsibilities have become more pressing.[58]

Japan's faith in the U.N. as the institution necessary to promote world security dates back to the 1950s. Signed in 1951, the U.S.-Japan Security Treaty was thought to be a temporary agreement that would exist only until the U.N. had established global security mechanisms.[59] Because the U.N. never developed a military force to ensure global security, an article which was an integral part of its original mission,[60] the Japanese opted to maintain the security arrangement with the United States.

The Japan Socialist Party which had gained some political momentum after World War II was most optimistic about the U.N.'s global security responsibility. For the Socialist Party, it was only with a strong U.N. which would protect Japanese independence that Japan could realize its hopes of economic growth and continued industrial development. Soon after the signing of the San Francisco Peace Treaty and the U.S.-Japan Mutual Security Treaty, the Japan Socialist Party became divided over whether or not to support these accords.

The left wing of the party endorsed 'unarmed neutrality' and accordingly opposed both treaties. It resisted capitalist *and* communist influence, desiring an independent Japan. The left wing of the party was deeply committed to the nation's peace constitution, had a commitment to its global diffusion,[61] and saw the U.N. as the institution capable of bringing about international cooperation and security. However, this wing of the party also realized that the U.N. was itself being divided because the United States and Soviet Union were striving for global dominance.[62]

The right wing of the Japan Socialist Party favored acceptance of both treaties. On the one hand, this wing of the party claimed to advocate neutrality for Japan; however, on the other hand, it made clear that communism does not accept the notion of neutrality. This wing of the party therefore maintained that while it wanted to see both security and independence for Japan guaranteed by the U.N., this was not about to happen in the short-term period, and therefore a peace accord with the allies and the security treaty with the United States, for the time being, was the only practical alternative.[63]

The position of the right wing of the Japan Socialist Party in some ways paralleled that of mainstream political thinking in Japan. Most Japanese policy makers continued to believe that the U.N. would assume its expected, international mission, but faced with the alternatives of Soviet domination, or no military protection, taking refuge under the U.S. security umbrella became the accepted course. Over time this evolved into the mostly American belief that because Japan received free protection from the United States, this at least partly explained its rapid economic growth and technological advancements. Although some Japanese have accepted this way of thinking,[64] most have interpreted U.S. pressures on Japan for increased defense spending as an effort to shift military responsibilities to a nation that has been committed to nonaggression. It is because of its pacifism that Japan chose economic assistance rather than military spending as a way to activate global leadership.[65]

In any case, Japan's desire to see the U.N. take on the responsibility of providing global security was genuine. It is this interest that suggests early signs of Japan's dissatisfaction with its paternalistic ties to the United States. U.N. security brought with it the likelihood of permitting an enlarged international role for Japan while simultaneously reducing its isolation. Thus for many Japanese a brighter political future appeared to be in store for their nation in 1956 when it was given membership to the U.N. It was also at this time when foreign minister Shigemitsu declared that, among other things, the U.N. should strengthen its efforts to bring about global peace and that it should work to end the existence of weapons intended to induce mass destruction.[66] Even though neither of these two points became reality during the Cold War, Japan still retained its faith that the U.N. would become a true international peacekeeping institution.

One year after being admitted to the U.N. Japan became a nonpermanent member of the security council (a position it has since occupied several times). Subsequent to this the number of responsible positions acquired by Japanese did increase within the U.N.; however, these positions typically did not permit Japan to exercise much political influence. As a result, generally Japan supported the United States within the U.N.[67]

In more recent times Japan's U.N. influence has not reflected its superpower position. The number of Japanese holding positions within the U.N. has been relatively low compared to other major nations. For example, despite larger financial contributions to the U.N. than France, Germany, and Russia (the Soviet Union in the past), Japan has had fewer people working in the U.N. than these nations.[68]

The post-Cold War has not diminished Japan's optimism concerning the potential of the U.N. On the contrary, Japan is now determined to have a more forceful political voice within this organization. To do this, however, Japan must accomplish several things. Two important objectives should be to increase the number of Japanese personnel working in the U.N. to an amount that reflects Japan's financial contributions, and perhaps more importantly, these diplomats must overcome the historic 'inferiority complex' which has been the hallmark of these individuals.[69] Japan's personnel increases in the U.N. could occur by using it financial contributions as a leverage for a larger diplomatic presence. Obtaining a permanent seat on the U.N. security council is currently an important objective of Japan and would, along with a larger diplomatic presence, expeditiously diminish feelings of inadequacy.

Becoming a permanent member of the security council would likely revitalize the debate in Japan concerning the extent to which it should participate in international peacekeeping operations. Even now some of Japan's policy makers favor more involvement by the nation's peacekeeping personnel than currently permitted by law. They feel that it is important for Japan to participate more extensively in international peacekeeping because this would increase the chance of obtaining a permanent seat on the security council.[70] As noted in Chapter 5, the peacekeeping law passed in 1992 restricts Japanese involvement in settling disputes in foreign nations. Among the restrictions which limit the Japanese peacekeeping force to noncombat activities are that it must be a neutral contingent, it can only enter a foreign nation if a cease fire is in effect, and should the cease fire end, Japanese personnel must leave the country.[71]

Because the Cold War stalemate within the security council between the United States and the Soviet Union is over, the U.N. now has the historic opportunity to fulfill its mission as a global peacemaking institution.[72] In June 1992 U.N. Secretary General Boutros-Ghali issued a report entitled *An Agenda for Peace* which makes clear that the world community currently has the rare opportunity to take advantage of the new international environment. In this report the secretary general discussed new ways to enhance world peace in the post-Cold War environment.[73] Japan has endorsed the secretary general's proposals; in his statement to the U.N. in September 1992, former foreign minister Michio Watanabe commented that he held the secretary general's plan 'in the highest regard.'[74]

Since international peacekeeping is expensive and because of broadening U.N. efforts in this area, a part of the secretary general's report proposed the immediate creation of a $50 million peacekeeping fund. In accordance with the belief that the U.N. should be providing global security, Japan has supported the creation of this fund.[75] In fact, Japan went one step further. Using former foreign minister Watanabe's statement to the 47th general assembly of the U.N. in September 1992 as a way to propose the idea, Japan made clear that it planned to introduce a resolution that would guarantee funding for the 'start-up stage' critical to international peacekeeping activities. Because of this resolution, which was approved by the end of the year, the U.N. created a $150 million peacekeeping fund.[76]

At the same time that the U.N. began redefining its security mission by seeking new peacekeeping and peacemaking measures, and as Japan made known its objective of having a bigger, more responsible part in these efforts, the United States gave its first sign (more were to follow; see Chapter 7) in the post-Cold War period that it was not pleased with the organization's expansionary plans. During an address before the U.N. in September 1993 president Clinton stated that: 'The United Nations simply cannot become engaged in every one of the world's conflicts. If the American people are to say yes to U.N. peacekeeping, the United Nations must know when to say no.'[77] However, not having the U.N. involved in all global conflicts is problematic for several reasons. First, the U.N. would be openly recoiling from its objective of promoting world peace. Secondly, avoiding some conflicts would leave the U.N. open to sharp criticisms since such action could easily trigger charges of selected biases that support the interests of the more powerful member nations. Finally, assuming the failure of other peacemaking efforts, the lack of swift intervention by the U.N. could too frequently mean either the escalation or prolongation of conflicts.

Should Japan obtain a permanent seat on the U.N. security council it will need to make unequivocal the kind of involvement it intends to have in peacekeeping activities. As a permanent member of the U.N. security council Japan's views are likely to be different at times from those of the United States. These differences would become especially evident if Japan takes the position of being a 'conscientious objector' in armed international peacekeeping activities,[78] or if it actively begins to promote global disarmament. As the global hegemon since World War II, the United States has avoided serious consideration of general and complete disarmament since the Kennedy Plan was submitted to the U.N. in the Spring of 1962.[79] The combination of Japan's pacifism and its historic feelings concerning the U.N.'s global responsibility currently raises the prospect that Japan will seek to reroute this organization, moving it in a direction that it was taking during the 1950s.

Shortly the after the founding of the U.N. in June 1945 the General Assembly created the Atomic Energy Commission. This commission's principal objective was to work toward the elimination of atomic weapons as well as other major weapons of mass destruction. In 1952 the U.N. created the Disarmament Commission, which was actually a merger between the Atomic Energy Commission and the Committee for Conventional Armaments formed in 1947. During the 1950s U.N. members had many serious discussions on disarmament. These discussions centered on comprehensive plans for both partial and general and complete disarmament. Soon after becoming a member of the U.N., Japan became involved with other nations in sponsoring resolutions that sought to restrict and eliminate the testing of nuclear weapons. In 1960 Japan was one of five nations that co-sponsored a resolution that focused on the issue of the nonproliferation of nuclear weapons. (The United States abstained in the vote on this resolution.)[80]

Of course, nearly four decades ago Japan could do little, if anything at all, to get the U.N. back on the disarmament track. However, things are now much different. The U.N.'s increased activities in recent years and its interest in further expansion obviously coincide with the end of the Cold War, as they do with Japan's emerging recognition of its growing global responsibility. As a global power having to take on more international responsibility and as a permanent member of the U.N. security council, Japan could begin the difficult task of revitalizing world interest in general and complete disarmament. For certain, as a permanent member of the security council Japan could be a very effective leader of global disarmament. However, given its increasing global influence, even at the present time Japan could still do more work to promote global disarmament.

Though having done little in the past, for some time Japan has felt that its unique pacifist constitution makes it responsible for advancing global peace.[81] The recent emphasis on the promotion of global peace in Japan's schools is suggestive of its awareness of the need to take action. As a major power, Japan can now actively begin to work on getting other nations to adopt its position on nuclear weapons. Major restrictions exist in Japan that prevent it from becoming a nuclear threat to the world. The 'Three Non-nuclear Principles' prohibits Japan from owning, producing, or introducing nuclear weapons. The Atomic Energy Law prohibits Japan from building or owning nuclear weapons. Moreover, since signing the Treaty on the Non-proliferation of Nuclear Weapons in 1976, Japan has strengthened the national ban on the production or purchase of nuclear weapons.[82] Most important for Japan as a global superpower is that it has officially adopted the diplomatic position of 'ultimately eliminating nuclear weapons.'[83]

Even prior to its acquisition of a permanent seat on the U.N. security council, Japan could strengthen and strictly enforce its existing military-spending restriction on developmental aid that it distributes to foreign nations. Japan could also stipulate that nations desiring to participate in its international S&T programs demonstrate how they will meet specific disarmament objectives, particularly in the area of research and development; moreover, participation should be contingent on the written agreement that outcomes from involvement in its programs will not be applied directly or indirectly to military projects.

Ready for Japan?

Becoming a global superpower really had not been on Japan's national agenda. From 1950 until the mid 1980s Japan, like other nations, did not anticipate the end of the Cold War. Rather, it concentrated entirely on strengthening industry and, accordingly, shirked global responsibilities. 'Benign indifference' characterized Japan's foreign policy.

More recently, Japan has become aware that it cannot shirk international responsibilities indefinitely. Although Japan is aware of this, it has still not consistently taken major steps in this direction. The reason for this is Japan's evident uncertainty in assuming leadership responsibilities on a global scale. Not ever having to perform as a global leader in the past contributes to this uncertainty. Another part of this uncertainty is the apprehension that aggressively pursuing significant global leadership responsibilities could put Japan in a confrontational position with the United States in matters relating to world environmental problems, international lending institutions and, as we will see in much greater detail in Chapter 7, global disarmament.

Should Japan fully assume a global leadership position in environmental matters and take the accompanying responsibilities seriously, some tension with the United States is very likely. The United States has never been at the forefront of environmental reform. Both government and industry in the United States have typically been slow in making needed changes to protect the environment. As the global leader at work on environmental problems, Japan could easily interpret U.S. procrastination as an indication of ill will toward the international community. Under present conditions, these opposing positions on the environment are likely to exacerbate existing problems relating to trade, Japan's investments in the United States, and the continuous American pressure for greater access to Japanese research and technologies.

It needs to be emphasized that the possession of S&T will become increasingly important not only to the U.S.-Japan relationship but also to global politics. Indeed, in two important ways S&T is already replacing military power as the defining characteristic of a global superpower. First, S&T intended for commercial use helps to improve a nation's competitive positions in international markets. Secondly, because dual-use technologies suggest the more advanced nature of commercial relative to military technologies, a nation that demonstrates well-rounded civilian competency markedly increases its global image.

To the extent that the possession of critical technologies and advanced science does not supplant the military power of a single nation in the post-Cold War period, foreign aid will fill at least some of the remaining void. The absence of a major military threat increases the international status and influence of Japan as the leading provider of foreign aid, now even beyond Asia. Additionally, the combination of Japan's growing international influence and its long-standing belief that the U.N. will become the international instrument which promotes and safeguards world peace introduces the auspicious prospect, as we will see in Chapter 7, that its relations with the United States can improve.

The need for an improved superpower relationship between Japan and the United States, one based on equally evolving industrial and S&T strengths, is important because this will stabilize it so that it can withstand strain. Beyond the continuation of some (manageable) bilateral tensions which are inevitable, this relationship will be subjected to strain in the likely event that industrializing nations become increasingly attracted to Japan as a model for growth. Japan's technological capability and industrial strength help to make it a development model for other nations. Moreover, Japan's demonstrated accomplishments in keeping unemployment low and making the threat of poverty negligible combine favorably with its success in maintaining good labor relations and with minimizing income disparities. Thus because Japan's brand of capitalism could very easily begin to attract nations desiring to build new or expand existing industrial infrastructures, especially if they are impressed by Japanese industry's capacity to export, a stabilized and solid relationship with the United States is considerably more desirable than one that is beleaguered by ongoing problems that cumulatively and consistently atrophy bilateral relations.

Notes to chapter 6

1. On the contrary, it was secret U.S. activities by the C.I.A. in the 1950s and 1960s which were intended to influence the direction of development that Japan would follow. To do this, the C.I.A. made use of financial payments amounting to millions of dollars to selected conservative leaders of Japan's U.S.-leaning Liberal Democratic Party. See 'C.I.A. Blocks Release of Documents on Japan,' *The New York Times*, February 15, 1995, p. A11.

2. See Clyde Prestowitz, *Trading Places* (New York: Basic Books, 1988), pp. 127-128; Giovanni Dosi, Laura D'Andrea Tyson, and John Zysman, 'Trade, Technologies, and Development: A Framework for Discussing Japan,' in *Politics and Productivity*, eds. Chalmers Johnson, Laura D'Andrea Tyson, and John Zysman (New York: Ballinger, 1989), pp. 4, 19; and Masaru Yoshitomi, 'Macroeconomic and Schumpetarian Features of Japanese Innovations in the 1980s,' in *Japan's Growing Technological Capabilities*, eds. Thomas Arrison, C. Fred Bergsten, Edward Graham, and Martha Caldwell Harris (Washington, D.C.: National Academy Press, 1992), pp. 99-105.

3. According to Eisuke Sakakibara, who had previously filled senior-level positions in Japan's Ministry of Finance, 'Peoplism is given concrete expression in the form of employee sovereignty within the corporation'; rather than leading to the 'polarization of capital and labor,' employee sovereignty, which itself became central to 'Japanese-style' management or 'familial management,' helps to promote 'the reduction of capitalist authority.' Sakakibara also writes that: 'Capitalistic elements have undeniably become more pronounced in the Japanese socioeconomic system over the past ten years or so, but they have not yet had any significant impact on this cardinal principle of peoplism.' See his *Beyond Capitalism: The Japanese Model of Market Economics* (Lanham, MD: University Press of America, 1993), especially pp. 3-9.

4. (Overview) 'The Educational System,' *Economic Eye: A Quarterly Digest of Views From Japan*, Spring 1993, p. 2; Ikuo Amano, 'Japanese Schooling in Historical Perspective,' *Economic Eye,* Spring 1993, pp. 6-7; and current (1995) data from Ministry of Education, 'Ratio of Persons Advanced to Schools of Higher Grade,' accessed on the World Wide Web.

5. See 'New Nagasaki Lesson is Peace,' *The Boston Globe*, August 10, 1993, p. 9.

6. See The Tsuneta Yano Memorial Society, ed. *Nippon: A Chartered*

History of Japan (Tokyo: Kokusei-Sha Corporation, 1992), p. 71.

7. Information on poverty and public assistance from the following sources: telephone conversation with a Japanese official in the economic section of the Japanese Embassy, Washington, D.C.; Japan Information Network, 'Recipients of Public Assistance (1983-1994),' accessed on the World Wide Web; and Council of Economic Advisers, *Economic Report of the President* (Washington, D.C.: United States Government Printing Office [USGPO], 1996), p. 314, accessed on the World Wide Web.

8. See Economic Planning Agency, *Economic Survey of Japan 1989-1990* (Tokyo: Ministry of Finance/Economic Planning Agency, 1991), p. 198, Table 3-1-8; and Douglas Ostrom, *Economic Equality in Japan*, Japan Economic Institute Report (Washington, D.C.: Japan Economic Institute, February 1, 1991).

9. 'What Do Japanese Executives Really Make,' *Business Week*, April 26, 1993, pp. 60-61.

10. See 'Number of Japanese Firms in the U.S. Decreases but Portion Showing Profits Increases,' *The Nikkei Weekly*, February 28, 1994, p. 3.

11. Ryutaro Komiya and Ryuhei Wakasugi, 'Japan's Foreign Direct Investment,' in *The Annals of the American Academy of Political and Social Science* (Newbury Park, CA: Sage, January 1991), p. 52; and James Jackson, 'Japan's Financial Stake in the United States: How Stable Is It?' in *Japan's Economic Challenge* (Washington, D.C.: USGPO, October 1990, Study Papers Submitted to the Joint Economic Committee, Congress of the United States), p. 196.

12. U.S. Department of Commerce, *Survey of Current Business*, July 1991, p. 86; U.S. Department of Commerce, *Survey of Current Business*, July 1994, p. 157.; and U.S. Department of Commerce, *Survey of Current Business*, July 1996, p. 130.

13. Data is from U.S. Department of Commerce, *Survey of Current Business*, May 1993, pp. 102-103.

14. Komiya and Wakasugi, 'Japan's Foreign Direct Investment,' pp. 45-61. Besides the United States, this explanation applies to European nations where Japan's foreign investments have increased in recent years.

15. Solomon Levine and Makoto Ohtsu, 'Transplanting Japanese Labor Relations,' in *The Annals of the American Academy of Political and Social Science*, January 1991, pp. 102-116.

16. Robert Perrucci, *Japanese Auto Transplants in the Heartland* (New York: Aldine De Gruyter, 1994), p. 123.

17. See U.S. Department of Commerce, *Survey of Current Business*, May

1993, p. 91.

18. Perrucci, *Japanese Auto Transplants*, pp. 5-7. Of the six Japanese auto transplants operating in the United States, only two are unionized.

19. Komiya and Wakasugi, 'Japan's Foreign Direct Investment,' p. 59.

20. According to the United Nations, *World Economic Survey 1991/92: A Reader* (New York: United Nations, 1991), p. 176: 'Econometric evidence for 'free-riding' would be a negative correlation between a country's military spending and the defence spending of the other members of the alliance or the leading power in the alliance.' Given that the United States is the incontrovertible leader of the alliance, the author's computations show that using real 1988 dollars for the years 1978-1988 the correlation coefficient for Japanese and American military spending is a *positive* .980486. Thus following this line of reasoning the available evidence suggests that Japan is *not* a 'free rider.' Moreover, Japanese military spending compared to that of the members of the North Atlantic Treaty Organization during this same period yields a *positive* correlation coefficient of .978734.

21. U.S. Department of Commerce, *U.S. Industrial Outlook 1993* (Washington, D.C.: USGPO, 1993), pp. 16-1 - 16-3. Japanese machine tool companies serve Japanese automobile transplants, which in recent years have invested heavily in the United States. On the matter of Japanese manufacturing transplants in the United States, including the auto industry, see Martin Kenney and Richard Florida, 'How Japanese Industry is Rebuilding the Rustbelt,' *Technology Review*, February/March 1991, pp. 25-33.

22. See Anthony DiFilippo, 'The National Machine Tool Partnership: Industry Savior or Band-Aid?' *The New Economy: The International Periodical on Economic Conversion and Disarmament*, Summer 1993, p. 7.

23. Data found in William Zeile, 'Merchandise Trade of U.S. Affiliates of Foreign Companies,' in U.S. Department of Commerce, *Survey of Current Business*, October 1993, pp. 52-65.

24. (Ministry of Foreign Affairs) Yasuhiro Nakasone, *Address by Yasuhiro Nakasone, Prime Minister*, National Press Club, Washington, D.C., May 1, 1987, pp. 4-5, 9.

25. See The Japan Times, *Defense of Japan, 1992* (originally published by the Ministry of Defense; translated by Japan Times; printed in Japan); also see 'Socialist Party's True Believers Rebel After Murayama Scraps Long-Held Ideals,' *The Nikkei Weekly*, August 29, 1994, p. 1.

26. Science and Technology Agency of Japan, 'Formal Statement of AEC Chairman,' *STA Today*, June 1993, p. 2.

27. See Toyoda Toshiyuki, 'Beyond the Mind-set of Deterrence to Genuine Disarmament,' *Japan Quarterly*, July-September 1992, p. 293; and United Nations Secretariat, *The United Nations and Disarmament: 1945-1965* (New York: United Nations, 1967), pp. 123-125.

28. See Michio Hashimoto, 'The Japanese Experience of Tackling Pollution,' *Japan Review of International Affairs*, Winter 1993, pp. 3-29; and Kazuhiro Ueta, 'The Lesson's of Japan's Environmental Policy: An Economist's Viewpoint,' *Japan Review of International Affairs*, Winter 1993, pp. 30-49.

29. Organization for Economic Co-operation and Development (OECD), *OECD Environmental Data: Compendium 1993* (Paris: OECD, 1993) p. 290.

30. Science and Technology Agency, *White Paper on Science and Technology* (Summary), (Tokyo: Prime Minister's Office, November 1994), p. 23.

31. See 'Scientists Confront Renewed Backlash on Global Warming,' *The New York Times*, September 14, 1993, pp. pp. C1-C6.

32. For the years 1981 through 1991, r is -.475 and R^2 is .226.

33. OECD, *OECD Environmental Data*, p. 35.

34. Former prime minister Kiichi Miyazawa's speech at the United Nations Conference on Environment and Development in Rio de Janeiro, June 13, 1992 can be found in Ministry of Finance, *Japan's ODA 1992* (Tokyo: Association for Promotion of International Cooperation, March 1993), pp. 199-200.

35. OECD data found in *Economic Eye*, Winter 1991, p. 22.

36. OECD, *OECD Environmental Data*, p. 294.

37. National Science Foundation, *Tokyo Report Memorandum*, Tokyo Office, January 20, 1994.

38. Hideo Tsuchiya, 'Looking to Japan for Leadership,' *Economic Eye*, Winter 1991, p. 21.

39. Ministry of Finance, *Japan's ODA 1992*, pp. 199-200.

40. Naoya Kitamura, 'Using Foreign Aid to Combat Global Warming,' *Economic Eye*, Winter 1991, p. 19.

41. Tsuchiya, 'Looking to Japan for Leadership,' p. 21.

42. Yutaka Kosai and Kenji Matsuyama, 'Japan's Economic Cooperation,' in *The Annals of the American Academy of Political and Social Science*, January 1991, pp. 67-69.

43. Isamu Miyazaki, 'Economic Prosperity or a Strong Military?' *Economic Eye*, March 1982, p. 7.

44. Larry Nowels, 'Japan's Foreign Aid Program,' in *Japan's Economic Challenge* (Washington, D.C.: USGPO, October 1990).

45. Ministry of Finance, *Japan's ODA 1992*, pp. 15, 72.

46. 'Socialists Make Their Mark with Low Defense Budget,' *The Nikkei Weekly*, August 1, 1994, pp. 1, 19; and Japan's ODA figure for 1995 from Ministry of Finance, accessed on the World Wide Web.

47. See Ministry of Finance, *Japan's ODA 1992*, pp. 15, 72; and Ministry of Finance, *Japan's ODA, Annual Report 1994* (Tokyo: Association for Promotion of International Cooperation, March 1995), pp. 247-248.

48. Eileen Doherty, 'Japan's Foreign Aid Policy: 1987 Update,' in *Japan Economic Institute, Japan: Exploring New Paths*, ed. Jon Choy (Washington, D.C.: Japan Economic Institute, 1988) p. 108.

49. See Ministry of Finance, *Japan's ODA 1992*, especially pp. 93 and 101; and Ministry of Finance, *Japan's ODA, Annual Report 1994,* 251-253.

50. The Ministry of Finance's annual ODA report provides documentation of Japan's growing financial contributions to multilateral institutions; see, for example, its *Japan's ODA, Annual Report 1994,* p. 158.

51. Eileen Doherty, 'Japan's Role in Multilateral Aid Organizations,' in *Japan Economic Institute, Japan: Exploring New Paths*, ed. Jon Choy (Washington, D.C.: Japan Economic Institute, 1988), pp. 112-116.

52. See Ministry of Finance, *Japan's ODA 1992*, pp. 118-121; and Ministry of Finance, *Japan's ODA, Annual Report 1994*, pp. 158-161.

53. See Ministry of Foreign Affairs, *Japan's ODA 1992*, pp. 22-24; and Ministry of Foreign Affairs, *Japan's ODA, Annual Report 1994*, pp. 107-111.

54. Nowels, 'Japan's Foreign Aid Program,' p. 404. A more detailed analysis of other aspects of Japan's ODA not discussed in this chapter can be found in Anthony DiFilippo, 'Japan's Foreign Aid Program and Discord with the United States' (paper presented at the Allied Social Science Associations Meeting, San Francisco, California, January 5, 1996).

55. Richard Cronin, 'A Japan Dominated Asia-Pacific Region?' in *Japan's Economic Challenge* (Washington, D.C.: USGPO, October 1990), pp. 484-493. Although Brunei is an ASEAN nation, Japanese foreign investments there have been very limited.

56. Atsushi Yamada, 'Open Markets vs. Economic Cooperation: U.S. and Asia at Odds over APEC Agenda,' *By the Way*, March/April 1995, pp. 4-7.

57. Cf. Yoichi Funabashi, 'Japan and America: Global Partners,' *Foreign*

Policy, Spring 1992, pp. 29-30; and C. Fred Bergsten, 'The Primacy of Economics,' *Foreign Policy*, Summer 1992, pp. 6-7, 10-11.

58. See Joseph Nye, 'Coping With Japan,' *Foreign Policy*, Winter 1992-93, pp. *97, 106.*

59. Yoichi Funabashi, 'Structural Defects in Tokyo's Foreign Policy,' *Economic* Eye, Summer 1993, p. 28.

60. See 'United Nations Charter, 1945,' in Edmund Jan Osmanczyk, *Encyclopedia of the United Nations and International Relations* (New York: Taylor and Francis, 1990), esp. pp. 946-948.

61. See Ishibashi Masashi, 'On Unarmed Neutrality,' *Japan Echo*, vol. 11, no. 1 (1984): 47-62.

62. Seiichi Katsumata, 'The Japan Socialist Party Should Oppose Both the San Francisco Peace Treaty and the Mutual Security Agreement,' *Journal of Social and Political Ideas in Japan*, April 1965, pp. 41-42. See also the introductory comments by the editor of this issue, pp. 38-40.

63. Kanae Hatano, 'In Rebuttal of Those Who Oppose Both the San Francisco Peace Treaty and the Mutual Security Agreement,' *Journal of Social and Political Ideas in Japan*, April 1965, pp. 43-44.

64. See, for example, Funabashi, 'Japan and America,' p. 26.

65. Cf. Miyazaki, 'Economic Prosperity or a Strong Military?' pp. 4-7.

66. See Yasushi Akashi, 'Problems of Japan's United Nations Diplomacy,' *Journal of Social and Political Ideas in Japan*, April 1966, p. 15.

67. Ibid., pp. 16-17.

68. Koya Azumi, 'The Future of U.S.-Japan Relations: A Peaceful Alternative,' *Business & the Contemporary World*, Winter 1992, p. 84; and Shigehiko Togo, 'The Human Element: Sharing the Burden of U.N. Support Requires More Human Involvement,' *By the Way*, July/August 1993, p. 14.

69. Kinhide Mushakoji, 'The Thought and Behavior of Japan's Diplomats,' *Journal of Social and Political Ideas in Japan*, April 1966, p. 20. Relating to World War II, an enemy phrase still exists as part of the U.N. charter.

70. See the Economist Intelligence Unit, *Country Report, 2nd Quarter 1993, Japan* (London: The Economist Intelligence Unit, 1993), pp. 12-13.

71. Barbara Wanner, *Japan Tested by Peacekeeping Responsibilities*, Japan Economic Institute Report (Washington, D.C.: Japan Economic Institute, July 9, 1993).

72. Tokuji Hosono, 'Blueprint for Peace: U.N. Seeks to Expand Functions to Meet Growing Needs,' *By The Way*, July/August, 1993, pp. 20-23.

73. *UN Chronicle*, September 1992, pp. 2-4.

74. Ministry of Foreign Affairs, *Diplomatic Bluebook 1992: Japan's Diplomatic Activities* (Tokyo: Ministry of Foreign Affairs, 1993), p. 436.

75. Hosono, 'Blueprint for Peace,' p. 23.

76. Ministry of Foreign Affairs, *Diplomatic Bluebook 1992*, pp. 55, 444.

77. Excerpts from president Clinton's U.N. speech in *The New York Times*, September 28, 1993, p. A16.

78. Azumi in his 'The Future of U.S.-Japan Relations,' uses the term 'conscientious objector' in regard to Japan's new role in the post-Cold War period.

79. See Anthony DiFilippo, *Arms Control Versus Disarmament*, Briefing Paper (Washington, D.C.: National Commission for Economic Conversion and Disarmament, May 1991).

80. See United Nations Secretariat, *The United Nations and Disarmament: 1945-1965*.

81. The United Nations Association of the USA and the Asia Pacific Association of Japan, *The Japanese-American Alliance* (New York: United Nations Association of the United States of America, 1983), p. 21.

82. See The Japan Times, *Defense of Japan, 1992*, pp. 66-67.

83. Ministry of Foreign Affairs, *Diplomatic Bluebook 1992*, p. 65; Ministry of Foreign Affairs, *Diplomatic Bluebook 1995*, accessed on the World Wide Web; and 'Background on Japan's Defense Policy,' *By the Way*, July/August 1994, p. 17.

7 Face to face: Japan's challenge to the United States

This final chapter provides a straightforward explanation of the ongoing tensions and conflicts that have been increasingly evident in U.S.-Japan relations. Suffice it to say here that the national security interests of the United States and Japan have been different for some time, and America's attempt to redirect more attention to the economic aspects of national security to improve its competitiveness remain connected to its hegemonic objectives. This chapter also argues that Japan's search for a new international position should include working to establish a global system that reflects its political ideals. This will require Japan to move much further away than it has up to this point from its economic interests. As a post-Cold War superpower, Japan must give considerably more attention to broadening its global responsibilities. Most importantly, by moving aggressively to initiate a global disarmament agenda, Japan will be internationalizing its domestic political ideals. At the same time, this will provide the United States with the opportunity to utilize more efficiently its national resources to improve industry competitiveness.

New world order: the constraints of hegemony

When president Bush declared that a 'new world order' had begun in early 1991 he was speaking about new U.S. efforts to maintain or impose international security. For sure, he and many others understood that improving relations between the United States and the Soviet Union, at least partly precipitated by the latter's inability to maintain the quickening pace of the arms race, relaxed superpower tensions within the United Nations (U.N.). This lead to the opportunity for pseudo-cooperative international security efforts with which the U.N. had not been familiar. The attention of U.S. policy makers, moreover, was refocused: away from the cataclysmic threat of nuclear holocaust, and toward real and prospective small, regional wars.[1] Thus conventional wisdom maintained that a new world order had taken shape as the old world order, that is, the Cold War, was fizzling out of existence. The demise of the Soviet Union served to corroborate this change in global politics.

What the enthusiasts of the new world order failed to reflect on was that the ideological parameters that had sustained the Cold War over the years were responsible for helping to exacerbate the inter-capitalist rivalry between Japan and the United States. Specifically, this rivalry was not part of the thinking that led to the emergence of a new world order. The failure to include this rivalry, that is, the mitigation of bilateral discord, as an integral part of the new world order indicated both the definitive preference of policy makers at the time to deal with the international security dimension connected to U.S. hegemony, and their refusal to acknowledge forthrightly that government had to have an important part in the industrial economy. Indeed, the existence of the latter condition meant that new and ambitious policies directed by government would not be forthcoming.

Several years before the proclamation was made that a new world order was unfolding U.S. policy makers had been contending with the harsh reality that the nation's competitiveness had been severely challenged, including in high technology industries,[2] and, of course, it was Japan that was largely responsible for this change. The irony here is that even though U.S. policy makers' understanding of national security had been changing incrementally during the 1980s and, as we shall see below, began to include more economic issues, this proved to be an inadequate catalyst for them to seize the opportunity and create a new comprehensive economic agenda for the new world order. In other words, despite the shedding of Cold War ideologies, the behaviors of the Cold War remained largely in place. American hegemony had been decisively built on military power, and the opportunity to intermittently exercise it was not about to be abandoned by U.S. policy makers.

For the Bush administration, like its predecessor, the Japan problem was an inter-capitalist one between long-time allies. This led to the widely held belief that solutions were achievable. As we shall see later in this chapter, the dominant, though not the only, perspective of Japan held by U.S. policy makers within the Reagan and the Bush administrations was that in time it would become more like America. This perspective was instrumental in keeping at least the rhetoric of free market capitalism alive during the 1980s. As we have seen in this work, however, in reality trade became more ostensibly 'managed,' and science and technology (S&T) issues became increasingly problematic. In short, tension between the United States and Japan steadily grew.

Unlike earlier administrations, the Clinton White House has painted a different picture of Japan. It is a picture that suggests that Japan is an 'equal partner,' but yet must be pressured into compliance through aggressive trade and related negotiations. Although there is some realization of Japan's real and potential global influence, nevertheless, it must understand that it has a place in the U.S.-led alliance. Like other nations, Japan must therefore accept U.S. hegemony, since it must continue to rely on America's unparalleled military power during the post-Cold War period.

Perhaps the most visible reminder of the Bush administration is that through the catchword new world order it became clear that America's military power would not be undermined in the post-Cold War period. While the Clinton administration has indicated that the domestic economy needs to be repaired, it has largely retained the military features of the Cold War. Thus it has not sufficiently developed programs that would enable the nation, including U.S. industry, to reach maximum competitive potential. Although the Clinton administration has given lip service to 'conversion' (industry transition from military to civilian economy), beneath the liberal veneer, however, lies the same hegemonic objective that U.S. policy makers have had for decades. This objective stands in the way of making significant structural changes in the domestic policy of the United States and therefore in improving substantially relations with Japan.

National security: then and now

For decades after World War II, the phrase national security in the United States was understood almost entirely within a military context. The majority of U.S. policy makers and analysts considered economic and military strength as very much interrelated, both based on a system of interchangeable parts. So confident were most observers of the durability of America's economic strength that they, simply put, took for granted U.S.

industry's ability to compete effectively in home and foreign markets. Concerning government-industry relations, this meant that the demands of the Cold War preoccupied U.S. policy makers.

Perhaps it was because this went on for too long or perhaps there was simply too much support for the rejection of government-industry cooperation beyond the defense sector, but whichever the case when policy makers realized that America's competitiveness was in jeopardy in the 1980s they were forced to accept change. This change was not unprecedented but it was different because it quickly became evident that there was a permanency attached to it. Japan's economic success, its working industrial policy, and its ability to apply the fruits of S&T -- the fact that still a good amount of it was originally American technology, notwithstanding -- to its industrial economy prodded many U.S. observers to wonder whether these initiatives would work in America. What began as an avalanche of speculation resulted in the federal government becoming somewhat more active in S&T development for the nondefense sector of the economy. To be sure, there was always much circumspection that accompanied the introduction of these initiatives during the 1980s, lest the impression be given that policy makers were unjustifiably meddling in the economy.

What all of this meant was that U.S. policy makers had to concern themselves more consciously with national security issues that related to the economy, in addition to the attention given to the military. Time had done what it ultimately does, it brought change -- but this change was very significant for government since the old meaning of national security had to be reformulated to address the economic dimensions of U.S. competitiveness.[3] This need to include a more visible economic dimension in the understanding of national security, however, did not mean that this took precedence over the thinking that shaped the new world order.

Thus despite considerable reluctance, republican policy makers in the Reagan and Bush administrations had to sanction a more activist government. This included the adoption of S&T programs that had civilian purposes and increased attention to the S&T developments occurring in Japan. The latter consideration meant that the United States began demanding symmetrical access to Japanese technologies. At the same time, more forceful steps were being taken in the area of bilateral trade with Japan.

The liberal tag of the Clinton White House has permitted it to abandon the closet activism of the preceding administrations and manifestly support programs aimed directly at the commercial sector. During the 1980s government programs that had the claimed potential of benefiting the commercial sector typically had a spin-off or later a dual-use justification attached to them. Although policy makers have not abandoned this strategy, government programs need not have this rationalization today to receive support. The expanded effort of Commerce's National Institute of Standards and Technology is a good example of government beginning a program without the dual-use justification. Recent efforts that make the defense laboratories accessible to private industry, on the other hand, retain the dual-use justification, though the prospects for overall success are exaggerated by policy makers.[4] In any case, the Clinton administration's efforts fall far short of what are needed to revitalize American industry.

The redefining of U.S. national security to include a significant increase in attention given to economic matters during the 1980s was not planned, nor was it a formalized process. Since the nation's relative economic decline was something that policy makers could not politically dismiss, no practical alternative existed for them. Slowly but steadily policy makers, including officials within the Department of Defense, had to address the economic and related S&T components of national security. For Defense Department officials, a weakened industrial sector could only be disastrous for the military's effectiveness. It is no accident, moreover, that the notion of dual-use technologies (those that have both civilian and military potential) appeared during the 1980s.

Because weakened U.S. competitiveness became the central economic issue of the 1980s, increasingly national security efforts were being targeted at ways to regain it. Restoring U.S. competitiveness had the attraction of killing two birds with one stone: revitalizing industry was both good for the economy and for the defense sector. We can mark president Reagan's appointing of a commission to study the problem of declining U.S. industrial competitiveness in the first half of the 1980s as the formal beginning of America's efforts to redefine national security. Government did not have a monopoly in this sustained effort to redefine national security. The appearance in the private sector, for example, of the influential Council on Competitiveness provided added momentum to this ongoing activity.

It is Japan that is largely responsible for precipitating the change in the way the United States has been reconsidering its national security objectives. As we have seen earlier in this work, economic growth built on a steadily evolving S&T complex has been the central and ongoing objective of

Japanese policy makers. Early in the post-war period the Japanese recognized that improving foreign technology and working diligently to develop a strong research base would provide a solid foundation to a first-rate manufacturing economy. Thus apart from its relatively modest Self Defense Forces, Japan's national security has been defined mainly in an economic sense, since it has not been directly linked to a strong military objective.

Despite American efforts to modify national security objectives, sustained economic and political problems between Japan and the United States caused this relationship to fester during the 1980s. So even before the Cold War began winding down, a *new* Cold War was in the making. This one did not require a nation to develop weapons and supporting technologies before its opponent. Rather, it was an economic war that for Japan meant structured industrial policies and the cooperation of management and labor; moreover, to a large extent it required that S&T be directed at commercial development so that markets could be secured. Still caught up in the Cold War, few people, save the Japanese, were paying much attention to the fact that the arms race was slowly giving way to the knowledge race, that economic warfare was a bigger threat to the national security of the United States than the Soviet Union. Because the United States was not willing to make the needed comprehensive policy changes, it had to move toward increased protectionism and managed trade.

In Japan during the 1980s confidence was quickly building. Much of this occurred because the Japanese had become proud of their economic accomplishments. This pride translated into a loosely knit form of nationalism.[5] It was this heightened nationalism that created a conflicting situation for the Japanese. Although grateful for past support from the United States, the Japanese at the same time took the position that America's economic problems were self inflicted. This eventually resulted in the Japanese increasingly showing resistance to American demands. Aware of America's determination to maintain the upper hand in the alliance, Japan's increased confidence and resistance to demands have recently become the political staples of its negotiating posture with the United States.

Interpretations of the transition in the global economy

Acknowledging Japan as an economic and S&T superpower has been unavoidable for more than a decade. More recently, Japan's awareness of the need for it to consider expanding its global responsibilities in the post-Cold War period has been a source of much discussion. This discussion has become more comprehensive and includes nonJapanese and often centers on Japan's interests and its international position relative to the United States.

The spectrum of explanations describing U.S.-Japan relations today is not particularly broad, but it is characterized by extremes. On one end there is the view that relations between Japan and the United States could, under very unusual conditions, lead to war.[6] On the other end of the spectrum is the view that Japan should adjust to a new global position by assuming more responsibilities, but it should still accept America as not only an ally, but a very powerful, military dominant one.[7] Those who advocate this view, in other words, understand that Japan can no longer avoid global responsibilities, but its new leadership position should be largely, if not entirely, supportive of the United States. In between the two extreme points are the positions of most analysts who, though they understand that some amount of tension will persist between the United States and Japan, see the relationship evolving toward superpower cooperation.

Those who advocate a cooperative bilateral relationship typically envision the mitigation of serious economic conflict -- a new form of peaceful coexistence. With the right amount of external pressure, *gaiatsu*, from the United States, Japan will adjust to its position as an economic and S&T superpower and eventually develop responsibilities that befit its global presence. Following this line of reasoning a convergence in global perspectives will occur between these two nations, and *Pax Consortia* will replace *Pax Americana*.[8] Some prefer to talk about the economic 'interdependence' of the two superpowers.[9] The gist of this vision is that global responsibilities will be 'shared' by the United States and Japan. For example, while the United States retains much of its military capability to ensure world order, Japan would become the humanitarian of the world, using its generous foreign aid program to fulfill responsibilities in this role.[10] As grand an opportunity as this and similar ones may be for Japan, it would still leave the United States in a dominant position. Moreover, as has been the case in recent years, this dominant position will continue to encourage American policy makers to expect economic compliance from Japan; if not, it will face severe unilateral consequences from the United States.

While there is plenty of room for cooperation between the United States and Japan, especially in the area of S&T during the post-Cold War period,[11] one should not be swept away by false optimism and the belief that understanding each other better will pave the way for relatively trouble-free technology relations.[12] As we have seen in Chapter 5, because Japanese corporations perform a large amount of advanced basic research, there will continue to be serious sticking points in bilateral relations in the area of S&T. There is also the crucial point that the continuing drive by the United States to dominate its overall bilateral relationship with Japan will not be especially conducive to maintaining solid S&T relations.

When the United States began to redefine national security to include increased concern for economic issues it did so because its weakened competitiveness was a direct threat to its hegemony. In other words, the changed efforts in protecting national security also changed the approach to maintaining hegemony. Given the underlying strength of U.S. industry during most of the Cold War, America's military power was the most visible way in which it remained a hegemon. Because of America's relative economic decline and the new understanding of national security, the old approach to maintaining hegemony required modification. Thus the relative decline of American industry meant that the United States could not maintain its hegemonic position unless it regained its competitiveness. While somewhat exaggerated, the United Nations Industrial Development Organization argues that in the early 1990s the international policy of the United States shifted from military security to global economic issues, for the ultimate purpose of strengthening its domestic economy.[13]

The main point here is that it will be exceedingly difficult for the United States and Japan to enter a truly cooperative arrangement to settle economic differences as long as American policy makers seek to maintain global hegemony. In addition to whatever remains of the appeal to past debts, America's military advantage will continue to be the main political source of pressure placed on Japan for the purpose of gaining compliance to the economic demands of the United States. This was an important part of the troubled relation between Japan and the United States during the 1980s; under present conditions, it will be even more problematic in the post-Cold War period, since the Japanese do not see the security threat that they did in the past.

Japan's emergent nationalism, which has already become evident in its demonstrable resistance to the United States on trade and technology issues, is likely to develop more vigorously now that Japan's perceived security threat from the Soviet Union no longer exists. This evolving attitude can be detected from a recent statement made by a high ranking Japanese official in the Ministry of Finance: 'Japanese history since World War II is a history of Americanization. It's time to think about post-Western civilization. We have to stop being afraid of conflicting with the U.S.'[14]

Economic conflicts can be managed, but only among equals -- true partners. Moreover, even though global responsibilities can be shared more or less to the satisfaction of both superpowers, there is still the big problem of which nation ultimately is in a more strategically powerful position. America's maintenance of a massive post-Cold War military system cannot be adequately explained without including its hegemonic objective. Because such a system seriously restricts the United States from fully tackling its economic problems, trade disputes with Japan will remain a critical factor that seriously disturbs bilateral relations.

A big problem in the past has been that the majority of U.S. policy makers and analysts took the position that military spending really did not create undue hardship for the economy. For them, there was insufficient proof available showing that military spending and the economic growth required to remain competitive were related empirically. Taking this hard-line, you-haven't-convinced-me approach, however, only applied to the U.S. economy. Most believed that the military burden proved to exact a toll on the Soviet economy; and the Japanese were often reminded of their 'free ride,' and how their competitive successes were dependent to some extent on the failure to spend enough on the military.

Most important for our purpose, however, is that the Japanese have officially pointed out that America's modest economic recovery from the recession ending in early 1991 has been restricted by its defense spending, and that it will continue to impede future growth.[15] Although it was several years ago that the Japanese became aware of the incompatibility between military spending and economic growth,[16] an explicit statement in a government publication to this effect definitely sheds light on the economic conflicts confronting the United States and Japan. Specifically, such an acknowledgment makes evident that the Japanese government holds the view that U.S. policy needs to be corrected if strong, sustained recovery is to occur. This makes it easier to understand why Japanese policy makers have become increasingly reluctant to concede to American trade demands: the failure of the United States to take the needed corrective (structural) steps to improve its economy should not be a price that Japan should have to pay in negotiations. Compounding this is that Japan's recession-induced problems in the early 1990s made government officials even more unyielding in trade negotiations.[17]

A related matter is Japan's uneasiness with the increasing support for what it is calling 'strategic trade policy.' According to Japan's Economic Planning Agency (EPA), strategic trade policy is government intervention in the market performance of high technology industries. Strategic trade policy, in other words, is managed trade. A nation adopting a strategic trade policy is taking aggressive protectionist steps to strengthen its high technology industries. EPA explicitly uses the U.S.-Japan semiconductor agreement as an example of strategic trade policy and points out that employing it has not strengthened the competitiveness of American-made dynamic random access memory (DRAM) products. EPA maintains that policy makers adopt a strategic trade policy which has a 'nationalistic character because their goal is to raise the welfare of their own country at the expense of other countries' welfare.'[18] To integrate EPA's argument with the one developed in this chapter, the adoption of strategic trade policy by U.S. policy makers is an attempt to bolster America's national security at the expense of Japan.

Emerging internationalism: relating to a hegemon

Japan currently is in a transition period. There is definitely a national awareness that it must assume a new international position. The problem is that the Japanese have not yet settled on exactly what that position will be. Whatever form it takes, it will not be one that readily yields to U.S. economic pressures.

Although there are many possibilities, it is extremely unlikely that Japan's new global position will include the type of nationalism that contributed to its involvement in World War II. Not only is this because of its defeat in the war and the subsequent absorption of pacifism into its culture, but also because decades of solid economic growth have taught the Japanese important lessons. Thus in the traditional meaning of hegemony where nations have had military power to support national goals, Japan does not desire to be a hegemon. In fact, it is correct to say that currently Japan does not want to fill a hegemonic role no matter how it is defined.[19] This, however, does not exempt it from being drawn into a very intense bilateral struggle.

Before addressing some of the problems related to Japan's emerging global role, it is necessary to review briefly its post-war international perspectives. During the early post-war years Japan had a limited international vision for itself.[20] Its central objective, as we have noted earlier in this book, was to rebuild the economy and develop a strong S&T base. Japan's constitution prevented active military internationalism. Advocating a neutralist foreign policy and 'unarmed neutrality,' the Japan Socialist Party, which emerged as the leading opposition party, became the watchdog that checked any enthusiasm for renewed militarism in Japan.[21] Because of the continued strength of the economy, the Japanese recognized during the 1970s that there was a need to become a more autonomous nation (that is, from the United States) and to develop a responsible international agenda that corresponded with its economic power.[22] The paternal alliance that the ruling Liberal Democratic Party (LDP) had with the United States, however, precluded genuine autonomy from emerging. What is more, the general sentiment among most policy makers in Japan was that since the security treaty with the United States needed to be preserved there were definite limitations to this autonomy. Nonetheless, Japan's realization that increased internationalism was necessary did not disappear, it was just that the Japanese were not quite sure what form it should take.

By the early 1980s Japan took initial steps toward developing an increased global position, although in reality more was said than actually accomplished at this time. Prime minister Nakasone was successful in getting authorization to transfer military-related technology to the United States, convinced many other policy makers that it was necessary to increase military spending, and made a case that the nation's peace constitution was not unchangeable.[23] But these were not steps that would contribute to Japan's autonomy, for they were taken largely to satisfy policy makers in the United States.

Despite the limitations, other important events took shape in the 1980s that created the momentum for Japan to seek an international position. Japan's vibrant economy and rapidly maturing S&T capabilities were convincing indicators that it had attained a level of industrial development that not only demanded close attention from the global community, but also required responsible action on its part. Thus while it is true that the strong economic growth of earlier decades helped to build Japanese confidence, it was fundamentally the establishment of a sophisticated S&T base that pushed Japan into the superpower arena.

In the meantime, driven by yen appreciation relative to major currencies Japan's foreign investments began to grow sharply, giving its multinationals a fresh look at the global economy. Japan's movement toward a new international position was also being propelled by its growing foreign aid and influence in multilateral institutions. Still another important factor was the realization that Japan had already learned the lesson that rapid industrial growth had an environmental price tag attached to it, and that it had a responsibility to other nations to help in this area.

While all of this together meant that Japan had come to understand that compliance with U.S. economic demands was not always in its national interests, there were two problems, connected with it creating an international position for itself. First, Japan did not specify a framework for a new international position. The ingredients were there, but Japanese policy makers did not know how -- or were reluctant -- to put them together. Secondly, Japan was still very much constrained by its security ties to the United States. With the elimination of the perceived threat from the Soviet Union, not only was the situation right to develop a new global position, but the only remaining reason for not doing this had been removed.

Still, the Japanese have not yet fully engaged this task. Despite its slow start in crafting a comprehensive international position, Japan's economic problems with the United States have grown more intense. The Japanese have been disturbed by the excessive Japan bashing in the past and they are particularly troubled by the Clinton administration's apparent willingness to take harsh unilateral action against their nation. Although the Japanese have not caved in under pressure, what they have not fully understood is that U.S. reactions are those of a nation trying desperately to retain its hegemonic position.

This means that Japan's understanding and pursuit of its national security interests are often at odds with the United States. As a capitalist nation, Japan's intentions to control markets and to protect private sector advanced research are indisputable. Japan also appears ready to establish an Asian trading bloc which is likely to be perceived by the United States as exclusionary. Moreover, to a large extent because of its substantial financial assistance to Asian nations, it is certainly influential in APEC (Asia Pacific Economic Cooperation). While the United States is a member of the APEC forum, so are several Asian nations which receive most of their foreign aid from Japan. These actions do not make Japan a hegemon, however; nor do they make it a nation attempting to establish itself as one. The line between hegemony and economic influence is not as fine as one would think.

For a nation to be an effective hegemon, three 'relative capability' components need to be present: (1) a strong economic foundation; (2) the possession of or access to the most advanced technology and science of the period; and (3) a dominant military system. Japan only possesses components one and two, and as argued here, is unlikely to try to develop the third (military) component. In the past, the United States had all three components. Because Japan now possesses the first two components, the United States has been very much involved in a sustained effort to maintain as much bilateral dominance as possible.

The post-Cold War period has changed not only America's relations with the former Soviet states, but also with Japan. It is true that problems facing the United States and Japan today can be viewed from different angles. There are major cultural differences, and these connect directly to business relations, including the treatment of labor. Japan and the United States have also understood national security differently. Aware of the reduced importance of the military component today, Japanese policy makers have few reasons to succumb to U.S. pressures on economic and related issues.

Early U.S. reactions to Japan

Japan's appearance as an definitive industrial power more than two decades ago was troubling to the United States, but not especially threatening at first to its global economic leadership. Although U.S. hegemony had been scarred by Vietnam, this did not threaten the condition of U.S.-Japan relations. American policy makers saw Japan only as an economic power that was quickly and proficiently mastering the application of industrial technology. Japan was viewed as an economic power, but a 'fragile' one.[24] In other words, Japan was still a force that could be reckoned with. That Japan could undermine America's national security was inconceivable at the time.

When U.S. policy makers did become aware of Japan's threat to America's national (economic) security, as we have pointed out, an initial response was to study America's competitiveness problem so that it could be corrected. The main U.S. reaction to Japan's growing 'unfair trade' behavior by the Reagan and then the Bush administrations was that through negotiation and with pressure it would comply with American standards. But at the same time the United States was eyeing Japanese technology, aware of its benefit to America's military system.[25] In this prime minister Nakasone, as well as other LDP policy makers, were willing players in the U.S.-led alliance.

In addition to constant pressure, which has been the dominant characteristic of the U.S. strategy toward Japan, patience was also necessary. From the U.S. perspective, there was a twofold need for patience. It was necessary because America had to recover fully from the deep recession of the early 1980s so that it could begin to deal with its competitiveness problem. Patience was also necessary because U.S. pressure on Japan intended to bring about compliance with America's trade demands was a stopgap measure, since with time Japan would adjust and become a reasonable partner. But patience did not pay off for the United States. Economic recovery and currency adjustments that drove the dollar down proved not to be significantly related to America's competitiveness problem and, despite the ongoing pressure on Japan, it was not coming around to the satisfaction of U.S. policy makers.

Because Japan was not making satisfactory progress in adjusting to its position as an economic superpower, U.S. policy makers decided that it needed a little help. The Structural Impediments Initiative (SII) talks that began at the end of the 1980s were designed for this purpose. The SII talks have been connected to the dialogue which speaks of the potential for 'convergence' through the structural reform of the Japanese and the U.S. economies which, in turn, would help resolve conflicts.[26] To the extent that convergence does or does not occur at some future time is beside the main point: from the beginning the SII talks were not a mutual undertaking designed to start a convergence process. At one point in the SII talks the United States was pressuring Japan to implement more than 240 economic changes which, as one Japanese political scientist points out, was analogous to the demands America made on Japan during the occupation period.[27] Not only were most of the obligations for structural reform placed on the shoulder's of policy makers in Japan but, according to a leading voice there, the SII discussions were 'initiated by Washington as a means of prying open Japanese markets without giving in to the growing calls in the U.S. for managed trade with Japan.'[28]

Challenging the dominant U.S. strategy during the 1980s that pressure and time would bring about Japanese compliance was the 'revisionist' position. One could argue that some of the early American revisionist position was a relatively liberal reaction toward Japan, given its sensitivity to the differences between Japan and the United States.[29] Revisionists stressed that there were institutional and cultural 'differences' between the political economies of Japan and the United States. Thus the revisionists introduced a sense of economic incongruity into U.S.-Japan relations.

The common characteristic of revisionist thinking is that an effort must be made to understand that Japan is different from the United States, and even other Western capitalist nations. It is this recognition of differences therefore that sets revisionist thinking apart from the more direct pressure-oriented strategy. From the revisionist perspective, progress in bilateral relations with Japan required an understanding that it is different. This sensitivity to Japan's differences increased the chances for successful negotiations between Japan and the United States.

Although a 'waiting frustration' in America resulted in interminable Japan bashing, proponents supporting both U.S. reactions to Japan had two things in common. The first was the widespread sentiment that, though an economic superpower, Japan was an ally that was beholden to the United States. Having a historical dependence in three key areas -- economic aid, technology transfers, and military security -- it was only a matter of time before Japan adjusted to Western standards. The other thing held in common was that neither the revisionists nor those that advocated the direct approach to bilateral relations, that is, those with less concern for and sensitivity to the 'differences,' anticipated that Japanese compliance was something that should not be taken for granted.

The passing of paternalism

The one word that best explains the view that many Japanese have of the United States is ambivalence. Surely the Japanese recognize the past debt to the United States. This debt perhaps best explains why some Japanese observers have been willing to yield to the leadership of the United States. But many other Japanese understand that it is this debt which U.S. policy makers and analysts have used in attempting to bring about compliance. What is more, for some within this group a paternal Japan is something that should never have developed, while for others it is now an anachronism that should be assigned to its rightful place in history. As a whole this group has come to the understanding that the national security interests of Japan have been at odds with those of the United States. There is, moreover, the sentiment that there is currently no reason why Japan should suffer the consequences of the macroeconomic shortcomings of American policy or have its domestic policy reflect U.S. interests.

The best known example of this has appeared in book form.[30] However, other examples of this exist, such as Japan's increased pertinaciousness in trade negotiations with the United States, as we have already indicated, and the 1993 dismissal of the director of the Defense Agency, Keisuke Nakanishi, for his public comments regarding changing the constitution.

Even though the Japanese have long faced pressure from the United States to assume a larger military responsibility, and despite the protracted dialogue on this matter promoted by the LDP, Nakanishi's public remark to alter Japan's constitution so that it could participate in global security missions nonetheless created an uproar. Although the socialists suffered major losses in the election that took place in 1993, they were the largest force within former prime minister Hosokawa's cabinet. Long opposed to a show of military force, as we have noted in this work, the socialists, along with other groups within the coalition, pressured prime minister Hosokawa for the resignation of the Defense Agency's director.[31]

Japan's matured resistance to pressures from the United States is the direct result of it concluding that many of America's economic problems have been self-inflicted. And as we shall see below, the revisionist reaction to Japan has aggravated bilateral discord, arguably strengthening the autonomy of Japanese decision making. It is also noteworthy that Japan's conflict with America, as some Japanese observers had anticipated, has now been carried to center stage in the global political economy. Japan's current enthusiasm concerning the WTO (World Trade Organization) exemplifies its strong belief that the resolution of bilateral trade problems should be in the context of this international institution. Interestingly, all of this has occurred during the time when it has been inaccurately proclaimed that the world has reached 'the end of history' because of the cessation of the struggle between capitalism and communism.

Differences become peculiarities

Although we have argued above that some of the revisionist criticism began largely as a relatively liberal reaction to the conflicts between the United States and Japan, this kind of response was ephemeral. Toward the end of the 1980s the revisionists clearly favored using a unilateral stick to gain the compliance of Japan. At the same time, the revisionists helped to hasten the emerging consensus that because of the improving relations between the United States and the Soviet Union, Japan had become a bigger threat to America's national security -- and thus its hegemonic global position -- than any other nation. It was at this time also that the Japanese became aware that the American public viewed the economic challenge coming from Japan as more threatening than the Soviet Union.[32]

The revisionist position evolved from stressing differences to emphasizing the 'peculiar' features of Japanese economic behavior. This angered some Japanese, who interpreted the notion of peculiarities as an additional political chasm in U.S.-Japan relations.[33] The suggestion of peculiarity carried with it the idea that Japan's adjustment to U.S. expectations was not forthcoming and was not likely to be achieved by past practices. Moreover, we should note that bolstering the revisionist peculiarity argument was that Japan did not actively participate in the Persian Gulf Crisis.

Although the containment policies of the Cold War were being put to rest, the revisionist position was at the forefront of shaping the idea that Japan needed to be contained.[34] Given that Americans had been awakened to their new Japanese rival, revisionists thus were hard at work attempting to develop a containment strategy for dealing with Japan. There is no way to quantify the extent to which revisionist thinking impacted on U.S. government pressure being unilaterally and more intensely directed at Japan during the past few years, but there certainly is evidence that such an approach has been used and threatened by American policy makers.

Targeting Japan

In May 1989 the U.S. government engaged article 301 of the Omnibus Trade and Competitiveness Act of 1988. This article is known as Super 301 because of its significantly adverse effects on an alleged violator of what the U.S. perceives as 'free trade.' The Japanese hold the opinion that Super 301 had been specifically formulated as a unilateral sanction intended for use against their country. The Japanese see this article as a stratagem intended to adjust markets to U.S. expectations.[35] They also interpret its implementation as a direct violation of international law.[36] Thus it is reasonable to argue that some of Japan's policy makers viewed the implementation of Super 301 as the foundation for a trade war, but that the lack of a strong international orientation compatible with its superpower position frustrated a substantive response.

The Bush administration implemented Super 301 in an effort to redress the unfair trade practices claimed to have been existing in three Japanese industries -- supercomputers, satellites, and lumber. Reflecting Japan's mounting dissatisfaction with unilateral trade measures, including the existence of anti-dumping duties, is the recent appearance of its unfair trade reports. In these publications Japan identifies both what it perceives as the principal unfair trading nations and the major problems in global trade relations. These publications have become increasingly critical of the United States and demonstrate Japan's growing determination to point directly to America's trade unfairness.

Published by the Industrial Structure Council, the authorized advisory organization to the Ministry of International Trade and Industry (MITI), the first unfair trade report which appeared in 1992 stated that the U.S. 'tendency to take unilateral action is regarded as an indication of the country's impatience with the ability of the GATT's dispute settlement procedures to resolve disputes over particular trade issues.' The second report which appeared about a year later emphasized the 'abusive application of anti-dumping measures' by the United States and its 'much greater degree of transparency in its anti-dumping investigations compared to other countries.' The 1993 report also stressed that: 'Basically, unilateral action is taken only by the United States.' More recently, the Japanese government has been especially critical of American policy makers' willingness to rely on a 'numerical target-based trade policy.'[37]

Upon learning of Japan's big trade surplus through October 1992, the United States responded with anti-dumping duties placed on steel and machine tools, among others.[38] A few months later, when American policy makers learned of Japan's trade surplus for all of 1992, which exceeded $100 billion, they suggested that the Japanese take steps to cut this surplus. Specifically, the suggestion was that Japan cut up to half of its trade surplus by the end of 1995 with the United States and the rest of the world.[39]

Relative to other nations, the United States has been more inclined to impose anti-dumping duties on Japan. This is interesting since American policy projects the impression that it is not Japan's exports to the United States, but rather the failure of its domestic market to accept sufficient imports (closed markets) that is the main problem today. As of December 1995, the United States had 39 anti-dumping duties imposed on Japanese goods. This is more than two thirds of the total (56) anti-dumping measures imposed on Japan -- over twice as many as all of the nations making up the European Community, and thirteen times as many as Canada. (See Table 7.1.) The imposition of anti-dumping duties on Japan by the United States, we should note, is not new. This U.S. response has deep roots, corresponding historically to the bilateral trade disputes that date back more than two decades. Between 1970 and early 1993 the United States was responsible for over half of all anti-dumping duties imposed on Japan. Moreover, between 1970 and 1995 the United States imposed 19 anti-dumping duties on Japanese goods for a period of ten or more years,

Table 7.1 Anti-dumping measures imposed on Japan, 1980-1995

Country	1980-1995	Lifted measures	Still in effect
United States	46 (51.1%)	7 (20.6%)	39 (69.6%)
European Community	30 (33.3%)	16 (47.1%)	14 (25%)
Canada	14 (15.6%)	11 (32.4%)	3 (5.4%)
Total	90	34	56

Source: Subcommittee on Unfair Trade Policies, 1996 Report on the WTO Consistency of Trade Policies by Major Trading Partners (Tokyo: MITI, 1996), p. 115.

compared to 3 for the entire European Community and none for Canada.[40]

More recently, the United States has resorted to using an intimidation and threat tactic on Japan as way of attempting to gain compliance in bilateral trade disputes. The Clinton administration has, for example, more than once threatened Japan with another dose of Super 301 if appropriate steps are not taken to make the expected adjustments. Former trade representative in the Clinton administration, Mickey Kantor, responded to a query regarding the possible use of Super 301 by stating: 'That would be an option,' and reflecting the revisionist position added that: 'The Japanese economy is different and has to be treated as such.'[41]

The Clinton administration appears willing to exercise a direct and forceful approach in dealing with Japan and to continue promulgating threatening rhetoric. Japanese policy makers resent this strategy and find it very offensive, since being 'equal partners' should not mean their nation takes a unilateral beating. But more importantly Japanese policy makers are learning that they can resist, and that groveling compliance is not part of what a superpower does when it feels that its national security is being placed in jeopardy.

This hard line approach in dealing with Japan is perhaps best expressed as trade belligerency. Suggested solutions to the Japan problem within the Clinton administration are to employ 'aggressive unilateralism,' and develop 'multilateral rules' that would be adjudicated by a 'neutral multilateral institution.'[42]

Especially in the 1990s, unilateral pressure being used by the United States on Japan to bring about trade compliance smacks of managed trade: setting quantitative targets for selected industries to 'open' Japanese markets is a fairly straightforward indicator of managed trade. Under these conditions, moreover, multilateral solutions to America's trade problems with Japan are very unlikely to favor the United States. The Japanese are cognizant of this. Early in 1994 one senior Japanese diplomat commented in Tokyo that: 'If the U.S. is really confident that Japan is unfair in specific trade practices, then it should file a complaint with the GATT, rather than judging what's wrong and taking [a] punitive step on its own. If the GATT finds Japan guilty, we are willing to accept whatever punishment.' Similarly, in the 1996 bilateral semiconductor dispute, MITI supported the position that the WTO settle the problem. [43]

Japan's evolving internationalism, particularly reflected in increasing foreign aid and foreign investments, has the effect of neutralizing the 'neutrality' of multilateral institutions that once favored the United States. The decided edge that the United States has had in the operations of multilateral institutions is no longer guaranteed as it was in the past. Given Japan's growing importance in the international political economy, it is very doubtful that multilateral solutions will be a viable, let alone the preferred, option for the United States.

The Clinton administration got its chance to apply its hard-line approach intended for Japan's compliance in early 1994. Operating on the anachronistic assumption that Japan would wait, play all of its cards, and then at the last minute fold, the Clinton White House became the first administration in the post World War II period to experience Japanese resistance in trade negotiations that ended in failure for the United States.

As persistently as the Clinton administration continued to demand quantitative shares of Japan's markets for American goods, some of its efforts reflecting what the Japanese have called strategic trade policy (e.g., in telecommunications), the Hosokawa government remained equally adamant by refusing to yield to American pressure. Since at least mid September 1993 the Japanese government had made it clear that it could not accept quantitative targets in trade negotiations with the United States. During a luncheon address to the Foreign Correspondents' Club of Japan at this time, former MITI minister Kumagai did not equivocate when speaking on the issue of relations with the United States. In his words: 'we cannot accept the idea of so-called numerical targets.'[44]

After a number of prior meetings between representatives from Japan and the United States, the bilateral negotiations associated with the Framework for a New Economic Partnership culminated in a stalemated Washington session taking place in mid February 1994 between president Clinton and then-prime minister Hosokawa. This created worldwide speculation that a trade war between the United States and Japan was imminent. The single most important catalyst for this speculation was the Clinton administration's quick response to the collapse of the trade negotiations with threats based on the policy framework of 'aggressive unilateralism.' The Clinton administration made it immediately clear that it was giving very serious consideration to imposing Super 301 and that another possible retaliatory action could be for the United States to drive up the value of the yen against the dollar to make Japanese goods less attractive to American consumers. Taking its first genuinely public step that suggested a nascent part of a new international posture, a 'different' Japan learned how to resist U.S. pressure, and responded to American threats with strong retaliatory threats of its own, leaving the clear impression that a trade war would not be in the best interests of either nation.[45] The experience and confidence obtained by Japan in the 1994 trade negotiations with the United States subsequently enabled it to withstand American pressure in the 1995 auto dispute and in the 1996 semiconductor controversy.

Temporary solutions to U.S.-Japan trade disputes should not be ruled out since they would help to maintain the semblance of an alliance which endured the many years of the Cold War. These kinds of solutions are likely to appear to have greater durability if both sides concede something to the other, and if the United States makes less visible its predilection for hegemonic negotiating. Short of success of this kind, Japan and the United States will simply have to tolerate each other's perceived shortcomings in the immediate future.

The reduced visibility of hegemonic negotiating by the United States will not make any less real the fact that it exists. As a result, Japan's political doggedness evident during trade negotiations taking place in recent years has become a standard that will persist under existing conditions. Current economic thought in Japan does not endorse caving in to U.S. pressures, but rather it supports creating a more flexible domestic economy that is somewhat more accommodating to foreign interests.[46] This, however, will do nothing to correct the inadequacies in U.S. economic and S&T policies, those which prevent the economy from competing satisfactorily with Japan.

The key point to make here is that under present conditions there will be serious and continuing trade problems between the United States and Japan. Because temporary solutions are not designed to deal with long-term structural problems, intense political conflicts can surface at virtually any time. Thus U.S.-Japan trade problems could conceivably become worse over time and become complicated by other very important issues connected to the clashing of the international perspectives of Japan and the United States. For sure there is no guarantee that, once Japan fully develops an international role, it will be in accord with U.S. interests.

Japan's new internationalism

The nationalism that had been evolving in Japan for two decades has showed signs of maturing in the 1990s. Having taken the first giant political step toward defining an international posture by refusing to comply with U.S. trade demands, Japanese policy makers must now address several important issues: (1) how far down the road they are willing to take Japan to ensure its economic (national) security; (2) the amount of global leadership responsibilities to be assumed by Japan; (3) the degree to which the formation and implementation of these leadership responsibilities will contravene U.S. interests; and (4) the reality that there are risks inescapably associated with efforts to accomplish both national and global objectives.

This last issue most definitely does not imply the possibility of a military encounter with the United States. What is being suggested is that should Japanese policy makers choose to affix visionary leadership to a responsible and comprehensive international posture, the chances for sustained conflict with the United States increase in the absence of major changes in American domestic policies. (Here the reference is specifically to policy changes which will enable the United States to bolster its ability to compete with Japan.) Should Japanese policy makers select this choice, which is the most reasonable and rational for a superpower that stands on the ideals of economic equity, environmental sensitivity, humanitarianism, and peace, they will become aware that Japan's responsibility to the global community must supersede its relationship with the United States.

As pointed out in earlier chapters, Japan's concern for the environment and its existing efforts in this area make it a strong candidate to assume more global responsibilities. Understand that the more responsibility that Japan assumes on international environmental matters, the greater the probability that it will be locked into a rather contentious dispute with the United States. For the most part, neither American policy makers nor U.S. businesses leaders have seriously committed themselves to a sustainable environment that includes safe energy alternatives, restricting hazardous waste, and

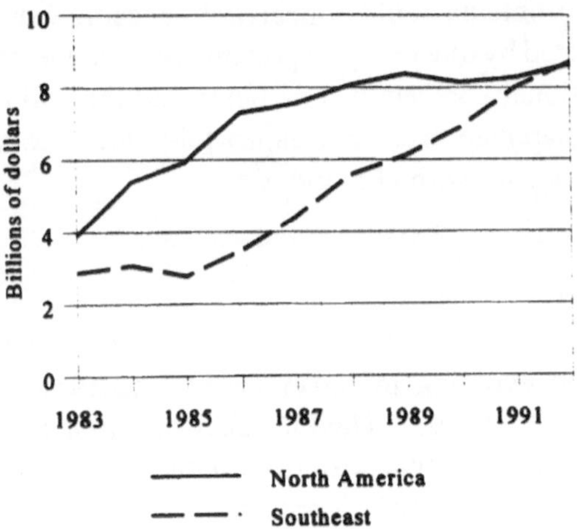

Figure 7.1 Japanese export trade to N. America and S.E. Asia, monthly averages, 1983-1992

Source: OECD, OECD Economic Surveys, 1992-1993, Japan.

developing technologies with improved efficiencies for future applications. The raising of environmental consciousness in the United States has been slow; foot dragging by both business and government has been more common than taking the initiatives necessary to improve the overall quality of the environment.

If Japan and the United Nations (U.N.) push for stricter environmental standards, and should the United States be indifferent to this, an additional wrinkle could easily surface in the bilateral relationship. Here it is imperative that Japan work in consonant with the U.N. This will help contain bilateral friction, since Japan will be relying on the legitimacy of the international community in seeking compliance to sound global environmental standards.

Japan's growing interest in the development of Asian nations, reflected in its expanding economic relationships with them, make 'neo-Asianism' an important part of its new international posture. The economic nurturing of Asia in recent years in the form of foreign aid and investments has been paying off for Japan. Figure 7.1 makes it indisputable that Japan's exports to what are known as DAEs (the dynamic Asian economies and China) have been growing steadily relative to the weakened trend evident for North America, of which the United States has been the principal destination area.

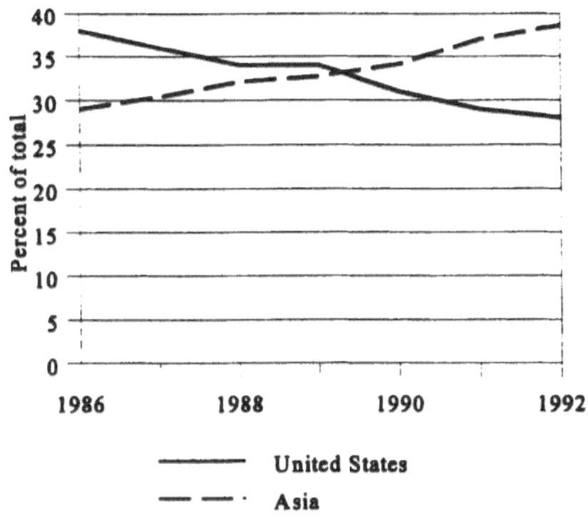

Figure 7.2 Japanese exports to the U.S. and Asia, 1986-1992

Source: Ministry of International Trade and Industry, Statistics on Japanese Industries, 1993.

During 1992 the average monthly amount of Japanese exports going to Southeast Asia had a dollar value ($8,698 million) greater than that of the export value for all of North America ($8,637 million). Considering Japan's aggregate export picture, while the percentage of its exports heading to the United States has declined precipitously since 1986, the exact opposite trend has been unfolding in the Asian nations. (See Figure 7.2.) Japan's exports to Asia currently exceed one third of its total; those to the United States amount to about 28 percent. Japan's exports to four the ASEAN nations -- Malaysia, Indonesia, Thailand, and the Philippines -- have made up over 20 percent of the total since 1991.[47]

Japan's growing economic ties to nations in Asia under present conditions are likely to create additional instability in America's Asian relationships for several important reasons. To the extent that these nations adopt the structural characteristics that had been 'peculiar' to the Japanese economy, U.S. trade relations with them are likely to become strained. America's trade problems with Japan, in other words, could spread and seriously include other Asian nations as their economies and technological infrastructures expand, encouraging them to develop more aggressive export policies.

Making this matter worse is that there is the very likely prospect that a yen-dominated Asian bloc will develop and that it will appear as exclusionary in the eyes of U.S. policy makers. Should the yen continue to ratchet upward relative to the dollar, Japanese manufacturers will more aggressively export to the Asian nations to help counteract reduced U.S. demand.[48] A yen-dominated Asian bloc will have the effect of alienating American business.

The attempt by Washington policy makers to talk up the value of the yen to dissuade consumers from purchasing Japan's products in retaliation for its failure to comply with U.S. trade demands is likely to have detrimental, long-run side effects for American producers. Such action will help to hasten and solidify a yen-dominated Asian bloc region by making it more attractive for Japanese producers to export there. Moreover, a higher yen will not only adversely affect U.S. investors in Japan but runs directly counter to the recent 'structural' adjustments made by Japan to facilitate foreign investment there.

Immediately subsequent to the conclusion of the first set of meetings that were part of the SII talks in June 1990 the government of Japan issued a 'Policy Statement on the Openness of Japanese Foreign Investment Policy.' Pointing to the importance of international investments to strengthen the global economy and to what it saw as the need to increase direct foreign investment in Japan, policy makers there introduced a number of structural reforms designed to do just that. These reforms included amending its Foreign Exchange and Foreign Trade Control Law, additional funds to JETRO (Japan External Trade Organization) to help promote more foreign investment in Japan and to the Japan Development Bank to provide low-interest loans to foreign investors, and the availability of tax breaks.[49] Guaranteeing these tax breaks to manufacturing companies that have more than one third of their capital base owned by a foreign enterprise was the enactment in 1992 of the Law on Extraordinary Measures for the Promotion of Imports and the Facilitation of Foreign Direct Investment in Japan.

There is the additional matter that as the Asian economies become more advanced and interrelated with Japan's, the threat of their initial accessibility to Japanese technology will irritate many U.S. policy makers and business leaders. As we have pointed out in Chapter 5, the United States has pressed Japan for access to its technology for some time. The possibility of technology being transferred to Asian nations knowing that this could jeopardize America's defense and commercial interests will not be well received by policy makers in the United States. This problem can easily become more complicated by increased transplants by Japan to the economies of its Asian neighbors and by co-production and joint venture operations.

Another area that is likely to become increasingly important to Japan's new international posture is S&T. Japanese programs that invite scientists and engineers from around the world to participate in international S&T projects are already welcomed by the global community. Increased funding and the continued proliferation of these programs will help remove the image of Japan as a closed society that has thrived off of the innovations of the United States, in particular. There are, however, serious problems in the area of S&T between the United States and Japan, as we have noted throughout this work. In order for both nations to work toward economic prosperity using their full potential, the possession of critical technology that is increasingly spawned from science is fundamental; for the United States, advanced technology is also indispensable for military systems.[50] Encouraged by government, Japan's private sector has already resisted U.S. pressures for access to advanced research work. The substantial amount of basic research now performed in Japan's private sector has the makings of a real trouble spot in the bilateral relationship.

In its quest for responsible internationalism Japan appears destined to increase its humanitarian efforts. Through its foreign aid program, Japan can assist many of the disadvantaged nations of the world, while it works to alleviate problems related to overpopulation in many nations. Largely through the Office of the U.N. High Commissioner for Refugees, Japan has recently increased its humanitarian aid to assist people in the erstwhile Yugoslavian region, and to help refugees in both Africa and Indonesia.[51] While work of this sort is highly commendable and should continue, as we have pointed out above, there is no sound reason why Japan should minimize its political and diplomatic potential in the changed international system because of the limited expectations of some of its domestic policy makers and observers in other nations.

The key point to be drawn from our analysis here is that there are a number of very serious issues that Japan must soon confront as it shapes its international position in the post-Cold War period. These issues connect directly to Japan's relations with the United States. The more Japanese policy makers refuse to yield under pressure, the more they venture out on an independent course for Japan. To the degree that they act on the recognition that preserving an alliance with the United States should not be the priority for their nation in the changed international environment, the more serious these issues become to the bilateral relationship. Our analysis, moreover, leads us to the conclusion that although the ongoing trade dispute between Japan and the United States is currently the most visible and the most serious problem in the bilateral relationship, it is not the only one.

That continued political discord can be avoided as the different facets of Japan's international posture evolves is possible, but increasingly unlikely, particularly since as things now stand the ongoing trade dispute, in some form or another, is not about to disappear. A superpower does not select political subservience as a way to deal with bilateral economic differences; it is more unlikely to do this when it has a commanding grip on many important markets, numerous industrial technologies, and is progressively moving toward establishing a S&T infrastructure second to none.

Global security and Japan

The single most important decision government officials in Japan must make regarding its new internationalism is the position they want the nation to take on global security. The need for this decision to be made surfaced during the Persian Gulf War. The national debate which continues, however, makes evident that Japan is not near finalizing its position.

There are two important impediments related to Japan's procrastination in making this decision on global security. The first is the generalized uncertainty among Japanese policy makers concerning the nation's diplomatic potential. The second is that compounded by pressure from outside of Japan to adjust to the standard of the new world order by militarily participating in international security missions, contravening political perspectives at home make it exceedingly difficult for Japanese policy makers to think through completely their nation's global position.

Japan's security alternatives

There are three major security alternatives available to Japanese policy makers, though only two of them have taken shape in the national debate. These alternatives for Japan are as follows: (1) change the constitution and permit the participation of its forces without restriction in international peacekeeping; (2) increase Japan's humanitarian energies to demonstrate concern for the security of the global community, and in the opinion of some who support this alternative, continue with restricted contributions to U.N. security missions; and (3) identify a new international security position that genuinely reflects Japan's post-World War II political ideals. (Should policy makers successfully sidestep the peace constitution without changing it and authorize the dispatch of troops abroad in a manner similar to what Western industrialized nations now do, this would then be analogous to alternative one.)

Advocates of the first alternative feel that it is mandatory that Japan amend its peace constitution to permit active participation by Japanese troops in international peacekeeping missions. As a major economic power, Japan can no longer avoid its responsibilities to the global community, because peacekeeping is a reality that the Japanese people must face up to at the present time. Several members of Japan's long-dominant LDP have supported this position,[52] particularly the faction led by Michio Watanabe. This position on constitutional change is definitely consistent with U.S. expectations of Japan.

To date, both unsuccessful and successful efforts to legitimate U.N. peacekeeping in the legislative process have faced the binding limitations of Japan's constitution. Former prime ministers Kaifu and Miyazawa's efforts to pass peacekeeping legislation were indicative of this.[53] Although calls for constitutional change have been heard in the past, the Japanese government's position on international peacekeeping still reflects the limitations of the existing constitution. Officially, the Japanese government maintains that: 'Japan, under the present Constitution, cannot participate in military actions the way that members of the North Atlantic Treaty Organization (NATO) can, although some in the United States and Europe hope for such Japanese participation. Given Japan's position in the international political arena, Japan should not do so either.'[54]

However, Japan's recent more determined search for an appropriate international position has given rise to inconsistency on the issue of constitutional change. To cite just one example, while former prime minister Miyazawa's foreign minister, Michio Watanabe, publicly called for a change in Japan's constitution to permit increased Japanese involvement in international peacekeeping missions, an LDP faction led by Yohei Kono opposed change.[55]

There are many problems associated with changing Japan's constitution to legitimize more active involvement in U.N. peacekeeping missions. To begin, constitutional change of this kind is a very big step for Japan since the initiation of the discussion is to a large extent the result of external pressure. Japan would be yielding to foreign pressure and in doing this would be relinquishing the opportunity to have a unique impact on world peace and disarmament -- two issues of great importance to the Japanese people. This kind of constitutional change also departs significantly from the behavior that has brought with it fairly good relations for Japan with its Asian neighbors, relations which have unquestionably been important to its economic interests.

Moreover, just as the existing constitution has met with attempts to provide a 'loose interpretation' to sanction international peacekeeping, a changed constitution would not be exempt from similar efforts and would be likely to invite even greater determination to expand Japan's military undertakings. For some time, there have been forces in Japan that have wanted to increase military spending above that which has actually taken place. This became recently evident during the spending authorization struggle for the 1994 fiscal year defense budget. While the Defense Agency pressed for an increase of nearly 2 percent in military spending, the Hosokawa government balked at this request and authorized less than one percent, the smallest increase since 1960. Former prime minister Tomiichi Murayama's 1995 military budget also demonstrated restraint in military spending.[56] The point is that with a changed constitution the debate over the limitations of Japan's military would be much more difficult to contain.

Equally threatening to the entrenched Japanese way of life is that S&T, so vital for present and future economic growth, would be absorbed for military contingencies to a greater extent than it is presently. With a changed constitution it would be difficult for Japan to avoid pressures from the United States to develop and share more 'peacekeeping' technologies to help America defray the costs of such an expensive enterprise. Thus changing Japan's constitution to permit more active involvement in U.N. peacekeeping missions would push the nation down a path much different from the one that has benefited it in the past.

One more important matter must be dealt with here. As a constitutional democracy Japanese policy makers need to fully consider the will of the people. Do the Japanese people want their nation to participate in international peacekeeping? While it is true that most people in Japan want to see their nation assume more global responsibilities, it is definitely not clear that active involvement in international peacekeeping and constitutional change are parts of this emerging global perspective.

Although almost 46 percent of the respondents to a public opinion poll performed by *Asahi Shimbun* at the end of May 1993 said that they supported the presence of Japanese peacekeeping personnel in Cambodia, a third expressed disapproval. Considering the very strict limitations placed on Japanese international peacekeeping personnel (see Chapter 6), it is more important to note that about two thirds of the respondents said that they felt uncomfortable with the government's increased activities in Cambodia and that 55 percent wanted to see Japanese personnel removed from Cambodia as expeditiously as possible, while less than a third supported following the plans of the U.N. Most significant of all, 60 percent felt that there should not be additional Japanese personnel sent abroad after Cambodia and the small contingent of Self Defense Forces in Mozambique. A survey by the *Nihon Keizai Shimbun* conducted in June 1993 showed that 50.5 percent of the respondents did not want the government to send Japanese officials abroad to support U.N. activities, while 42.5 percent were in favor of this.[57]

The Japanese public has similar reservations on the issue of constitutional change. A 1993 public opinion survey conducted by the Japan Broadcasting Corporation showed that almost 50 percent of the respondents opposed constitutional change to permit Japanese involvement in U.N. peacekeeping activities, compared to 38 percent that supported this.[58] In *Yomiuri Shimbun* surveys conducted in 1981, 1986, and 1991, public opinion was against constitutional change. In surveys conducted in 1993 and 1994, however, public opinion swung in the other direction. For example, in the 1994 survey 44 percent of the respondents favored constitutional change, while 40 percent opposed it. But in the 1994 survey the proportion of respondents supporting constitutional change declined significantly (by 6 percent) from 1993.[59] The enthusiasm and the subsequent decline in support for constitutional change that took place between 1993 and 1994, can be explained by the reduced international pressure on Japan (most of which came from the United States), and by the public uneasiness associated with Japan's troop presence in Cambodia. Specifically, when the effects from the pressure brought on by the Gulf War (the passage of the June 1992 Peacekeeping Bill) subsided and the uneasiness created by Japan's involvement in Cambodia appeared, public opinion toward constitutional change by 1994 began to move back to where it was in earlier surveys.

Supporters of the second security alternative see Japan as largely if not entirely making a civilian-humanitarian contribution to the global community. However, proponents of this alternative hold somewhat divergent views. Some do not envision constitutional change occurring in the immediate future. For them it is important to keep in mind the nation's past, and so utmost circumspection must be exercised at all times on the matter of international peacekeeping. When needed, Japan should provide 'peripheral' support to U.N. international peacekeeping. That Japanese forces are now permitted by law to participate in international peacekeeping is a reality they accept, but do not necessarily like. Others supporting this alternative reject constitutional change and the underlying motivation that Japan should participate in international peacekeeping missions like other Western industrialized nations. Moreover, this school of thought believes that the legislated change permitting limited Japanese participation in U.N. peacekeeping missions should be rescinded since it is unconstitutional. Certainly this is the position of part of the Socialist Democratic Party as well as some others in Japan.[60]

Perhaps more than the first alternative, by heading down this second path, as it is currently, Japan is placing itself in a position in which it is much more likely to follow the lead of the United States on international security matters. On the other hand, were constitutional change to occur (the first alternative) Japan could conceivably have more latitude in making the decision of whether or not to participate in peacekeeping operations. Given Japan's post-war pacifism and because the lives of Japanese forces would be put in jeopardy, policy makers would need to consider the prospect of public backlash more so than they would were the nation's personnel involved in providing peripheral activities. In any case, neither of these security alternatives can change the current post-Cold War reality that has left the United States as the only global hegemon. Thus similar to the new American security models that fully anticipate U.S. leadership,[61] the debate in Japan concerning which security alternative is best for the nation relies on exactly the same expectation.

Selecting either of these security alternatives puts Japan in a most precarious position. Although the trade problem and the ever looming possibility of a trade war is the most obvious disturbance in Japan-U.S. relations, there are other formidable problems, some of which currently relate to S&T issues. By selecting either of the above alternatives, both of which assume U.S. leadership, Japan curtails its influence in international security matters and in virtually all bilateral issues. While this may be fine for Western industrialized nations, it would not be good for Japan. For sure, Japan is not like other industrialized nations. Not only is its lingering trade surplus with America troubling to observers, but its advanced research and technological capabilities are broadly comparable only with the United States.

The growing importance of the economic dimension of national security for the United States most definitely increases the chances that it will continue to use its leadership in international policy making to influence economic relations with Japan. The 'new Cold War' -- the intensified intercapitalist competition that most prominently involves the United States and Japan and that centers directly on market control during a period when many national economies face the prospect of slow growth, and which at a more fundamental level includes the requirement to possess leading edge S&T -- demands that foreign policy shift ostensibly to economic issues. Without attempting to regain at least some of its lost international competitiveness, the effectiveness of U.S. foreign policy and thus its hegemonic stability is adversely affected.

It is in this context that we should reflect on the following statements appearing in President Clinton's 1995 budget message: 'We have put our economic competitiveness at the heart of our foreign policy, as we must in a global economy. We are following the success of NAFTA and GATT with further market-opening negotiations and the intensified focus on the promotion of U.S. exports. We are paying particular attention to the Asian and Pacific markets, which have the most dynamic growth of any region of the world.'[62]

That the United States would use its foreign policy and specifically its leadership position in international security as a leverage to strengthen its competitiveness should be expected. Although Japanese policy makers may desire a relationship among equals, such can not be the case as long as the United States maintains its exclusive leadership position in international security. When trade disputes surface Japan will always be susceptible to making some concessionary settlements, despite the desire of policy makers to project a different impression. Similarly, the United States will continue to press Japan for access to its S&T, claiming its necessity for international security. Japan's concessions to the United States in the past can easily be tied to its perceived need for security. Thus should Japan now select either of the two security alternatives described above, concessions to the United States will be the result of its political uncertainty created by the desire to define for itself an international role that corresponds with its superpower position and its failure to act in a manner that most accurately reflects its post-war security ideals.

The third security alternative available to Japan requires an ongoing effort that will enable it to create a distinguished international position for itself. This alternative would permit Japan to work cooperatively with the global community and remove the dependency constraints formed by U.S. hegemony. But most importantly this third alternative would give Japan's policy makers the opportunity to define an international position that parallels the political values embedded in its peace constitution and which reflect the dominant sentiment of its population.

In a survey of Japan's adult population prepared for the prime minister's office in 1989 nearly two thirds of the respondents mentioned the 'existence of the peace constitution' as one of the factors (among eight) that helped to account for how Japan maintained peace since World War II.[63] This was by far the most important of the multiple factors given by the respondents, almost 20 percent more than the second most common response, 'stability of Japan's economy and society,' and over 20 percent more than the third most popular response, 'Japan-U.S. security arrangements.' Significantly, as the 'single most important factor' identified by the respondents, almost 40 percent mentioned the peace constitution, more than double the amount that felt that Japan's strong economy and society had this effect.

Compared to an earlier poll conducted in 1981, the 1989 prime minister's survey showed some interesting changes. It indicated that beyond defense policy, over 60 percent of the respondents expressed the view that the 'positive promotion of peace diplomacy' should be greatly expanded, an increase of 8 percentage points from the 1981 survey. If we include the percentage of those who felt that diplomacy needs to be strengthened to some extent, then over 80 percent expressed the need for Japan to improve its diplomatic activities. The most commonly identified way in which Japan should improve its diplomatic efforts, indicated by over 43 percent of the respondents, is to rely more heavily on the U.N., an increase of nearly 8 percentage points from the 1981 survey.[64] Additionally, the Japanese people have expressed increasing optimism concerning the effects of their government's disarmament efforts in helping to bring about global peace; nearly two thirds of the respondents felt this way in the more recent survey, compared to about 46 percent in 1981. Also reflecting the growing optimism of the Japanese concerning global disarmament is that compared to the 1981 survey where about 30 percent of the respondents were hopeful about its realization, the more recent survey indicated that 60 percent held this expectation.

A *Yomiuri*/Gallup poll conducted in 1994 showed that 50 percent of the surveyed Japanese respondents felt that the U.N.'s major objective should be to stop and resolve regional disputes. Japanese support was considerably higher than that found in the United States, Great Britain, Germany, and France. This survey indicates that the Japanese are much more likely than people in these other nations to see the U.N. -- and not the United States -- as responsible for policing the globe.[65]

A public opinion poll conducted by the Japanese government in 1993 indicated the following points. Showing an increase from 1989, nearly 46 percent of those surveyed agreed with the statement that 'Japan, which has become a major country, has the responsibility toward the international community to promote its own internationalization.' This public opinion poll also showed that nearly half of the respondents felt that Japan should make a 'contribution to solving global problems,' and almost 30 percent believed that Japan should make a 'contribution to international peace, including arbitration of regional conflicts,' both categories markedly increasing from 1989.[66]

While it is true that public opinion is always susceptible to change by one force or another, the survey data reveals that the people of Japan believe strongly in peace diplomacy, see the U.N. as an extremely important agent in this process, and generally see global disarmament as a real possibility. Perhaps this reasoning in conjunction with the nearly five-decade-long commitment to the peace constitution helps explain why the Japanese people feel that their nation's diplomatic efforts have contributed to world peace. This popular sentiment, however, stands in stark contrast to the initiatives undertaken by Japanese government in promoting global disarmament.

A nonstellar disarmament record

Japan does not have a strong historical record in the U.N. on disarmament-related issues. Generally, Japan has followed the lead of the United States when voting on U.N. resolutions involving the control of weapons and disarmament, or it has decided to abstain. Thus for the most part Japan's post-war diplomacy has not been exercised independently.[67] Unfortunately, through the Cold War Japan had to contend with diplomatic ambivalence relating to disarmament initiatives because of its commitment to U.S. foreign policy.[68]

Despite the fact that many Japanese policy makers saw their nation's unique peace constitution and other pacifist legal restrictions such as the Three Nonnuclear Principles as providing them with the incentives to make an important and leading contribution to global disarmament,[69] Cold War politics suppressed their actions. Policy makers in the LDP accepted both protection from the United States and its deterrence strategy. Thus Japan's disarmament ideals became a catastrophe of the Cold War. In 1979 Japan voted against the U.N. resolution that would forestall any additional placement of nuclear weapons on the soil of nations not already possessing them. Japan's official explanation for this vote was that in addition to verification being a problem, supporting this resolution could have meant tipping the balance of global power and this, in turn, would have unsettled international peace and security. Included also in the Japanese position was the point that nations possessing nuclear weapons had to proceed gradually toward the goal of global disarmament.[70]

This deterrence strategy could still be observed in 1991 as the Cold War was drawing to a close. Japan voted against a proposal for a 'nuclear-arms freeze' even though the core paragraph of this resolution specifically requested the United States and what was then the Soviet Union 'to reach agreement on an immediate nuclear-arms freeze.' In explaining its vote against this resolution, Japan once again pointed to the verification problem, held that deterrence was still necessary to ensure global peace, and argued that a freeze would keep alive the fear of nuclear superiority, whether real or not, and so would have no effect on global disarmament.[71]

As we noted earlier in this work, for years Japan refused to support proposals prohibiting the use nuclear weapons. Moreover, for several years, it did not vote in favor of resolutions prohibiting the manufacture of new weapons of mass destruction.[72] As recently as 1992, Japan abstained from the resolution on the 'amendment of the treaty banning nuclear weapons tests in the atmosphere, in outer space and underwater.'[73] While U.N. resolutions voted on by the general assembly relating to international security are not binding, but are considered only recommendations or invitations to act,[74] Japan has definitely not taken much of an independent initiative in this area (broadly defined) either. For example, between 1986 and 1992 Japan was not the chief sponsor of any U.N. resolution on disarmament related-issues.

Some recent examples of Japan's disarmament diplomacy

Despite its poor performance in the past on disarmament-related issues, there is a distinct possibility that Japan's diplomatic efforts in the post-Cold War period will move from the back seat of the global policy machine steered by the United States to a dynamic, independent position. Perhaps the least indicative of this transition but nonetheless worthy of mention is that in 1992 Japan changed its opposition vote from the previous year on the

U.N. resolution concerned with the nuclear-arms freeze, deciding that an abstention more appropriately reflected its position.[75]

With the Netherlands, Japan did introduce in November 1991 the U.N. resolution 'transparency in armaments.' Eventually supported by 40 nations, but not the United States, this resolution in its final version sought to establish a 'conventional arms' register to scrutinize arms transfers and to create a mechanism that would help account for the possession of weapons by nations.[76] Moreover, Japan is committed to a comprehensive nuclear test ban. It has supported a U.N. decision made in August 1993 which would authorize negotiations on a treaty concerned with the prohibition of all nuclear testing by the end 1996.[77] Japan's interest in the ban on nuclear testing is not new, however. In fact, at the Conference on Disarmament in 1984 it was Japan's foreign minister who tried to negotiate a settlement that centered on the problem of verification by proposing a 'step-by-step formula' so that agreement could be reached on the nuclear test ban.[78] For years, Japan has maintained that such a ban would constitute the initial step in global disarmament.[79]

Japan has officially taken the position that it is only natural that its foreign policy be directed at disarmament-related issues, and that in doing this it needs 'to assume a leading role in this field,' relying on what it calls 'cooperation for peace.' Welcomed by much of the international community, Japan recommended improvements in the International Atomic Energy Agency in 1991.[80] In September 1993 Japan advanced a plan for 'international monitoring' to the International Atomic Energy Agency. One of the key objectives in its proposal was to stop the spread of plutonium as Russia and the United States disengage weapons systems.[81] Since 1989 Japan has annually hosted the U.N. Conference on Disarmament Issues, a clear indication of its concern for global disarmament.[82]

The removal from power of the LDP in 1993 created some additional momentum for a disarmament regime in Japan. One important change related to the political composition of prime minister Morihiro Hosokawa's cabinet. Although widely disparaged because of their appreciable loss of seats in the lower house of the Diet in the 1993 election, members of the Social Democratic Party of Japan (SDPJ) dominated prime minister Hosokawa's coalition government. The SDPJ has had a long history of supporting disarmament, and accordingly made a commitment to advance it while in office; moreover, at the time, the party continued to reject the constitutionally of Japan's Self Defense Forces.[83] The SDPJ also opposed Japan's new international security activities in the 1990s from the start, made it perfectly clear that it was against any further expansion, and remained committed to restraint in the military budget.

In addition to the disarmament interests of the socialists and, most significantly, as we have seen above, those of the people of Japan, former prime minister Hosokawa himself was a supporter of this important issue. Only a few months after taking office he pointed out that Japan should 'take the initiative' in reducing weapons throughout the world.[84] Besides crafting the smallest defense budget increase in decades, Hosokawa's relationship with the Defense Agency had been further soured by his plans to revise Japan's military posture. In line with the position of members of the SDPJ, Hosokawa wanted to reduce the number of Self Defense Forces.[85]

Still other events offer reason for sober optimism concerning Japan and global disarmament. In 1994 Japan introduced to the 49th U.N. General Assembly the draft resolution 'Nuclear Disarmament with a View to the Ultimate Elimination of Nuclear Weapons,' a resolution subsequently adopted by most member nations. Japan has also vehemently opposed the recent resumption of nuclear testing by both China and France and was one of a number nations supporting the adopted U.N. resolution calling for the cessation of nuclear testing.[86] At the grassroots level, in early December 1993 the Article Nine Society of Japan (A9SJ), along with two other groups, signed a petition in Tokyo requesting that the U.N. incorporate the principles of Article 9 from the Japanese constitution (which renounces war) into the U.N. charter. A9SJ has approximately 4,000 members, and reportedly there are about 100 other existing organizations in Japan that endorse its views.[87]

Toward a nonhegemonic world order: Japan's place

One word sufficiently characterizes Japan's economic history after World War II; that word is growth. As we have argued in this work, Japan's economic growth has occurred in large part because of the undivided attention that the nation gave to developing first its technology and subsequently its science base. Japan did what any nation would do if it genuinely wanted to build its economy, its industries, and improve the quality of life for its people. Thus investments and time were put into what mattered most. But when the effects of Japan's success began to cross paths with the United States, problems emerged. Trade friction with the United States, which gave its best to winning the arms race, increased steadily, and since the 1980s exponentially. Stunned by the Japanese challenge to its national security and eventually realizing the potential diminution to its hegemonic position, the United States reacted by intimating to Japan what it had done for the country. But the 'free riding, paternalism effect,' had limitations, and this meant that economic sanctions continually had to be imposed on Japan to ensure compliance. Although the Clinton administration has recently boasted that improvement in the bilateral relationship has been made through American pressure and that Japan therefore is manageable with the right leadership, during the 1990s relations have proven thus far to be troubled. If Japan is truly going to make a major mark on the post-Cold War order it has to move beyond its preoccupation with the correct way of handling the bilateral relationship. It must adopt a position of 'responsible global leadership.'

While aware of its superpower potential and its responsibility to the post-Cold War international community, Japan nonetheless has not yet found its place. On the one hand, Japan wants to show cooperation with the United States. On the other hand, there is the emerging recognition that a superpower must not be a follower, and must not continue to give in to the United States. Although the Japanese have learned this lesson, they have not fully understood what the responsibilities are of a true superpower.

Japan still sees its security relationship with the United States as an important pillar of its defense and as significant to the bilateral alliance. In the post-Cold War period most especially, this is directly contradictory to Japan's claimed support for a much stronger and reformed U.N., assuming that it has not abandoned its historical expectations of this organization (see Chapter 6). By failing to promote the U.N. as *the* global security organization and therefore responsible for its protection, what the Japanese forty years ago labeled 'U.N.-centered diplomacy,'[88] Japan continues to lend credence to the arguments of those who contend that its foreign policy has been passive and focused on narrow economic interests. Japan's narrow economically driven national interests were indeed formed during the Cold War and, with the U.S. security relationship, largely comprised what it considered to be its national security. However, as a global superpower in the post-Cold War period, Japan's must insist on changes that correspond to a new international position for itself.

Japan has approached this problem both tentatively and awkwardly. Thus Japan has been preoccupied with being a team player on global security matters and has not taken substantive steps to strengthen the U.N. as the world's security organization. Since the crisis in the Gulf, Japan has been far too concerned with a peacekeeping responsibility like a 'normal' nation -- this emerging mainly because of outside pressure -- and has avoided addressing the issue of collective security *completely* under U.N. aegis. Like other industrialized nations, Japan has accepted U.S. military power and thus its leadership in the U.N. Consequently, aside from policy-related statements,[89] Japan has had to forego its long-held belief in the U.N. as the international security body, and this, in turn, has constrained its efforts in the area of global disarmament.

A strong U.N. that endorses, practices, and manages collective security[90] devoid of hegemonic influence is not at all incompatible with global disarmament. The compatibility of a collective security system completely controlled by the U.N. and disarmament was specifically addressed by the Japanese representative to the Eighteen Nation Disarmament Committee in 1969. After referring to the preamble of the Japanese constitution which recognizes the inherent right of all people throughout the world to live without the fear of war, the Japanese representative stressed that permanent peace 'cannot be realized unless the peace-keeping operations of the United Nations are strengthened to such an extent that each nation will be able to give up its own armaments and to rely entirely for its security upon the United Nations collective security system.'[91] Today both the disarmament of individual nation states and collective security completely under U.N. management remain very much interrelated: combining both of these objectives eliminates hegemonic influence, precludes the possibility of regional military power struggles between nations, and most obviously it frees up resources, thereby permitting countries to decide how and where to apply them in a nonmilitary fashion.

In its search for a new international position it would certainly not be to the detriment of Japan to work assiduously to achieve these objectives. Attempting to achieve these objectives will permit Japan to assume an international position that truly reflects its constitution, as well as other legal standards such as the Three Nonnuclear Principles. Most importantly, these objectives are not at all incompatible with the interests of a sizable proportion of Japan's population. Although pursuing these objectives are not likely to improve Japan's relations with the United States right away, any resulting political problems need to be tolerated and ultimately, as we will see below, should be partially healed by the economic opportunities becoming available to the United States.

In the short-term period, however, Japan faces an uphill struggle with the United States on this issue. President Clinton now rejects the idea of an international force under U.N. control, notwithstanding the fact that as a presidential candidate he endorsed such an entity; moreover, there is currently some pressure in Congress to abandon the little progress made by the U.N. during the 1990s.[92] The recent retrograde action by the United States to move away from a U.N. international security paradigm in the post-Cold War period provides Japan with a unique chance to step in and assume a leadership position designed to strengthen this organization.

Not unlike the past, perhaps one of the most problematic issues attached to disarmament today is verification. Nations, including the United States and Japan, have failed to support U.N. resolutions on many occasions in the past because they have argued that verification would be a problem. We have spent a considerable amount of time in this work emphasizing that Japan's S&T base has become increasingly sophisticated since World War II. An innovative Japan determined to hold an important position in the post-Cold War era could begin serious work on developing verification technologies that would expedite the disarmament process. Not only is Japan financially equipped to begin such an undertaking but it also possesses the personnel to implement such programs.

New verification technologies would be very important to global disarmament. Japan already is deeply involved in seismological research.[93] By extending this research to interrelate with its commitment to the ban on nuclear testing, Japan could make an important contribution to improving verification technologies, thus taking a critical step in the direction of global disarmament.

Also stressed in this work is that one of Japan's recent contributions to the global community has been the introduction of international S&T programs. Japan could begin to support international S&T programs that would address the problem of verification. These programs could be linked to its domestic work in S&T.

Japan has recently awarded a sum of $20 million to the International Science and Technology Center to stop the exodus of scientists and engineers from the nations that were formerly part of the Soviet Union and who had been involved in building weapons of mass destruction.[94] There is no reason why these scientists and engineers could not be fully integrated into international programs designed to tackle verification problems. Moreover, Japan could help allay the U.S. criticism of it receiving a 'free ride' by attracting Americans to these programs also. One of the biggest problems today facing American engineers especially, but scientists as well, is job loss resulting from reductions in U.S. military spending. In addition to relaxing the eligibility requirements for U.S. engineers and scientists who need retraining to perform civilian work,[95] Japanese-funded international programs focusing on verification problems could absorb some of these occupationally displaced Americans.

Japan's foreign aid program, which as we have already seen includes formal restrictions on assistance to nations too deeply involved in military efforts, needs more definitive criteria so that Japanese government officials do not have as much latitude as they do presently in the distribution of funds. Although the new principles that are part of Japan's foreign assistance effort are commendable, both the lack of specific criteria for the distribution of aid and the fact that the program itself is too closely linked to U.S. security interests are problematic. By establishing and making use of precise criteria for aid distribution and relating them to the military efforts of nations, Japan's foreign assistance program could become a very important instrument in the promotion of global disarmament. Those nations taking the initiative to reduce their military budgets and to disengage and dispose of weapons could be placed on a priority list, enabling them to acquire aid from Japan on a more timely basis.

While Japan is beginning to show some interest in taking on leadership responsibilities to address problems in the global environment, it can do much more in this important area. Taking substantive steps in this area will bolster Japan's confidence. Such steps will also demonstrate to other nations that Japan can be an effective leader when it comes to an important international problem.

Japan also needs to press harder within the U.N. for increased responsibility, specifically by acquiring a permanent seat on the security council. Having the second biggest economy in the world and as the second largest financial contributor to the U.N., Japan's absence from the security council makes no sense. The fact that Japan was asked by the U.N. in May 1996 to pay its $128 million in unpaid dues early, primarily because the United States was in arrears,[96] begs the question of it holding a permanent seat on the security council and indicates the potential influence it could have in reshaping this organization. Even though it has received support from the United States for a few years, Japan has not been aggressive enough on the matter of acquiring a permanent seat on the security council.

As a permanent member of the U.N.'s security council, Japan would be in a position to be much more influential in the area of global disarmament. What is more, Japan has received the assurance of U.N. secretary general Boutros Boutros-Ghali that additional participation in peacekeeping activities is not a requirement for becoming a permanent member of the security council.[97]

What about U.S.-Japan relations?

The most important question to emerge from the above discussion concerning Japan adopting an international position that promotes collective security and global disarmament is how these efforts will affect its relations with the United States. Can the bilateral relationship survive these Japanese initiatives? The implementation of these policies, while creating an increased shakiness in the bilateral relationship in the short term, will give way to adjustments and eventually acceptance. This is so for two important reasons. The first is that the economies of both superpowers are very much interdependent. Related to this is that there are physical limitations, in both an economic and political sense, to responses by the United States. Secondly, a major benefit to the United States, should it plan to allocate its national resources, including S&T, more efficiently while developing supportive domestic policies, will be the gradual restoration of its industrial competitiveness. The United States will no longer have to rely on government pressure, at least not to the extent that it does now, or wait for market failures stemming from recession in Japan, to realign industry and improve competitiveness.

Still, as the remaining hegemonic power, it is not likely that the United States will welcome Japanese initiatives dealing with disarmament and collective security under U.N. management because such action, to the extent that it is successful, will steadily reduce America's global influence. For example, such action by the Japanese will obviously underscore the irrelevancy of the U.S.-Japan security relationship. With this Cold War relic removed from the bilateral relationship, tensions could very well mount on important issues relating to trade differences and problems in S&T that have plagued Japan- U.S. relations for some time.

Thus the response by the United States to the Japanese initiatives outlined above is likely to be initially characterized by some asperity based on what it considers to be political naivety. U.S. hegemony has thrived off of the security uncertainties in the world in the past, and this has not changed significantly today. However, if bilateral relations are going to improve, the Japanese need to take steps to apply pressure on the United States that will provide it with the opportunity to become more competitive.

Ever since the Japanese challenge to the United States became evident, American policy makers were quick to point to Japan's low military budget as an important reason contributing to its success. This gave rise to an ongoing discussion which characterized Japan as being in the unenviable position of accepting a free ride. Perhaps because of the reluctance to envision the United States without its military might, the majority of policy makers and other observers would not reason that the nation's overextension in this area had the reverse effect, that is, that America's competitiveness had been partly drained by its long-term and continuously massive military investment.[98] Even in those fleeting moments when such ideas crept into policy statements, it became easier to use 'scapegoat economics' on Japan than to risk the end of American hegemony caused by undermining the military system.

This is not to say that in the past Japan has never transgressed the principles of fair trade, nor should it be exonerated for infrequently moving beyond its parochial national interests. For indeed, throughout the Cold War Japan's foreign policy has been quite narrow and focused largely on supporting its economic objectives. To the extent that these transgressions have existed, however, they pale in comparison to the effects from the almost entirely different S&T paths selected by the United States and Japan. That trade differences in what increasingly became technology- and science-based industries emerged as the most critical in the bilateral dialogue and that problems relating to the exchange of S&T have also appeared should be absolutely no surprise. Much of the antagonism that has emerged in U.S.-Japan relations can be likened to what physicists call the 'Doppler effect': coupled with other important strengths, including good management skills and worker-employer relations, an undiminished work ethic, and unfailing attention to quality improvement, as Japan's S&T infrastructure became more advanced yielding more competitive industries, many American businesses, lobbying organizations, policy makers and other observers steadily increased the level of 'noise' concerning the unfairness of Japan and its unwillingness to cooperate fully with the United States.[99] Thus U.S. pressure on Japan to change, in one form or another, became more and more evident.

Certainly the U.S.-Japan semiconductor dispute can be explained in this way. Having lost its dominant semiconductor market share and technological leadership to Japanese producers, American manufacturers, joined by policy makers, continued to complain about Japan's unfair business practices. The position of the United States in 1987 that Japan's domestic market had been largely closed to its products, that is, relative to the European market, prompted American policy makers and some business officials to demand increasing access. While the preference for Japanese manufacturers to consume domestically made chips may have been strong, such consumption patterns hardly 'prove' the American contention of closed markets. European semiconductor producers are not by any stretch of the imagination comparable to the many world class manufacturers in Japan. As is generally known, largely because of U.S. pressure the number of American-produced chips purchased by Japan has increased dramatically in the past few years. But what is not widely understood is that to placate the United States, 'administrative guidance' in Japan literally had the effect of forcing Japanese semiconductor consumers to purchase what have been considered poorer-quality and higher-priced chips than those produced by domestic firms.[100]

For years Japanese semiconductor producers such as Hitachi, NEC, Toshiba, and Fujitsu spent heavily on research, development, and equipment. The recession in the first part of the 1990s, however, took its toll on Japanese semiconductor producers by cutting into expenditures for research and capital investments. In addition to the recession in Japan helping U.S. manufacturers regain some lost competitiveness, it provided South Korean chip makers with the opportunity to surge in the global market. The growing share of the Japanese semiconductor market currently held by U.S. manufacturers and their improved competitive position notwithstanding, American pressure on Japan continues to exist. Even though the foreign share of the Japanese semiconductor market reached 30 percent in 1996, heated discussions over several months continued on the issue of whether or not to renew the bilateral semiconductor agreement.

Unless Japan begins to apply pressure on the United States, situations like this are inevitable. U.S. industry leaders have little to lose and everything to gain when they work to persuade government to act on their behalf. With precedents already existing in a number of industries, and with the Japanese seen as willing to use unfair tactics and as not prepared to cooperate fully, American business leaders have the momentum of the past on their side. To change this, Japan must do things differently. It is therefore no longer good enough for Japan to resist pressure from the United States, as it has done since 1994. As a superpower Japan must initiate some pressure of its own on the United States at the same time that it works to influence the global system.

Why Japan should act

The most important issue is not whether or not Japan will act, but that it *should* act. Japan's 'differences' would be most usefully applied in the manner outlined here. Indeed, the suggestion for Japan's new international security responsibility presents it with a very big challenge. Still, if Japan were to embrace our proposal with the determination to implement it similar to what it did when it sought economic growth and technological advance in the past, it could be successful. It is true that what we have proposed is a fundamentally different challenge than what Japan has confronted for many years. It is, nonetheless, one that will help strengthen Japan as much as it will other nations around the globe. In the past few years there is some evidence that Japan's interests in the U.N. and disarmament have been somewhat revitalized; clearly much more is needed to achieve success.

Because Japan had been led for decades by policy makers aligned with the LDP, a party whose perception of global politics largely mirrored that of the United States, this will not be an easy task. It is also the case that for decades Japan has had little experience in developing independent international initiatives that directly effect the nations of the world. Critics of Japan's minimal diplomacy do not leave much room for it to develop a true leadership agenda. Western analysts are not reluctant to show their pessimism concerning the appearance of an active leadership position for Japan in today's global political economy.[101]

However, these critics neglect to point out that the past policies of the LDP to some extent have been discredited in Japan. The LDP, moreover, no longer has a guaranteed, monopolizing hold on Japanese politics, as it did in the past. Much more important than this is that the present period creates the prospect for taking down the political sign posts that directed the Western world and Japan for decades and replacing them with entirely new ones. It is up to Japan to decide if it wants to assume the leadership responsibilities that will provide different direction and thus help create a different global system. If Japan is truly seeking a unique international position it will not become contented with environmental and humanitarian responsibilities, as commendable as these are. Its national value structure and the sentiment of its people can together become the political cornerstone of its new international position.

Japan has frequently pointed to the huge budget deficit of the United States as a very important structural problem that must be confronted by American policy makers. However, since the United States still performs its Cold War policing activities around the globe,[102] this has certainly been a major obstacle preventing it from substantially reducing the budget deficit. Similarly, still far too much of America's S&T remains tied up in its global security policy. By applying pressure on the United States, both directly and through the U.N., and thus engaging the active support of other nations, Japan could begin to stress that it is now America's chance to get a 'free ride.' Accompanied by a planned and verifiable disarmament process, collective security managed by the U.N. would give the United States the chance to redirect its national resources to more constructive endeavors.

At the same time, Japan would find itself occupying a very important international position by advocating and taking substantive steps toward the implementation of disarmament and collective security. Working to acquire the cooperation of other nations would be an important part of Japan's new international position. To date, however, the Japanese have not come to the realization that this unique international position is directly related to relieving the ongoing pressure that the United States has applied on Japan.

Under present conditions, the long-evident bilateral scuffling that has centered on trade and S&T, will continue, and perhaps worsen, no matter what changes are made by Japan. Very easily these problems could be extended to include serious differences concerning the environment, should Japan's global interests move in this area. The end of the perceived Soviet threat thrusts issues relating to political economy on to center stage and leaves little room for political pressure to subside. Without the introduction of significant change, Japan-U.S. friction will remain the central and permanent national security struggle between these two superpowers. With American hegemony intact, the United States will always be in a position to place additional pressure on Japan to change, including in matters relating to global security.

Thus as a post-Cold War superpower Japan must give serious thought to changing the script that has prevailed over the international system and dominated the U.S.-Japan relationship for many years. Accepting assigned responsibilities from the United States will not permit Japan to experience the rewards connected to its attaining a superpower position, since these would preclude the complete realization of objectives that reflect the values of its people and its constitution. Nor will these assigned responsibilities and concessions to the United States do anything to reduce the friction evident in the bilateral relationship.

Notes to chapter 7

1. Murray Weidenbaum, *Small Wars, Big Defense* (New York: Oxford University Press, 1992).

2. Anthony DiFilippo, *From Industry to Arms: The Political Economy of High Technology* (Westport, CT: Greenwood, 1990).

3. See Joseph Romm, *Defining National Security: The Nonmilitary Aspects* (New York: Council on Foreign Relations Press, 1993) who points out that national security is an ambiguous phrase, with different meanings and subject to change; and Michael Borrus and John Zysman, 'Industrial Competitiveness and American National Security,' in *The Highest Stakes*, eds. Wayne Sandholtz et al. (New York: Oxford University Press, 1992), pp. 7-52.

4. Anthony DiFilippo, 'The National Machine Tool Partnership: Industry Savior or Band-Aid?' *The New Economy: The International Periodical on Economic Conversion and Disarmament*, Summer 1993, p. 7.

5. Kenneth Pyle, *The Japanese Question: Power and Purpose in a New Era* (Washington, D.C.: American Enterprise Institute Press, 1992), p. 65.

6. George Friedman and Meredith LeBard, *The Coming War with Japan* (New York: St. Martin's Press, 1991).

7. See, for example, Kitaoka Shin'ichi, 'Opting for a Global Alliance,' *Japan Echo*, vol. 19 (special issue), 1992, pp. 26-36; and Stephen Bosworth, 'The United States and Asia,' *Foreign Affairs*, vol. 71, no. 1, 1992, pp. 113-129.

8. Cf. Fred Bergsten and Paula Stern, 'A New Vision for the United States-Japan Economic Relations,' in *Harness the Rising Sun: An American Strategy for Managing Japan's Rise as a Global Power*, eds. Joseph Nye and David Rowe (Lanham, MD: University Press of America, 1993), pp. 99-116.

9. See *U.S.-Japan Interdependencies, Hearings Before the Joint Economic Committee, Congress of the United States*, December 17 and 18, 1991 (Washington, D.C.: U.S. Government Printing Office [USGPO], 1992).

10. Edward Lincoln, *Japan's New Global Role* (Washington, D.C.: The Brookings Institution, 1993), p. 242.

11. Anthony DiFilippo, 'Working with Japan to Revitalize Industry,' *IEEE Technology and Society Magazine*, Winter 1993, pp. 21-27.

12. Lewis Branscomb, 'U.S.-Japan Technology Relations: The Road from Resentment to Understanding,' *The Harvard Journal of World Affairs*, Spring 1992, pp. 103-115.

13. United Nations Industrial Development Organization, *Industry and Development, Global Report 1993/94* (Vienna: United Nations Industrial Development Organization, 1993), p. 1.

14. Quoted in 'Emergence of Neo-Asianism Reflects Japan's Identity Search,' *The Nikkei Weekly*, January 17, 1994, p. 22.

15. Economic Planning Agency, *White Paper on the World Economy 1993* (Summary), December 19, 1993, p. 5. According to this report, defense spending is one of the 5 d's that is moderating economic growth in the United States; the other four are debt, downsizing, deficit, and demography.

16. Isamu Miyazaki, 'Economic Prosperity or a Strong Military?' *Economic Eye: A Quarterly Digest of Views from Japan*, March 1982, pp. 4-7.

17. See 'Neither Side Budges in Framework Talks,' *The Nikkei Weekly*, January 17, 1994, p. 2.

18. Economic Planning Agency, *White Paper 1993*, pp. 18, 22, 25-26; and 'Economic Planners Rap U.S. for Strategic Trade Policies,' *The Nikkei Weekly*, December 13, 1993, pp. 1, 27.

19. Koji Taira, 'Japan, An Imminent Hegemon?' in *The Annals of the American Academy of Political and Social Science* (Newbury Park, CA: Sage, January 1991), pp. 151-163.

20. Pyle, *The Japanese Question*, pp. 20-41.

21. For details see J. A. A. Stockwin, *The Japan Socialist Party and Neutralism* (London: Melbourne University Press, 1968). In 1991 the Japan Socialist Party changed its name to the Social Democratic Party of Japan.

22. Kosaka Masataka, 'Japanese Americans in a Competitive Alliance,' in *The Silent Power: Japan's Identity and World Role*, ed. Japan Center for International Exchange (Tokyo: Simul Press, 1976), pp. 167-188.

23. See Pyle, *The Japanese Question*, pp. 101-103.

24. Zbigniew Brzezinski, *The Fragile Blossom* (New York: Harper & Row, 1972); and Frank Gibney, *Japan: The Fragile Superpower* (New York: W.W. Norton, 1975).

25. See 'For Perry, The Best Defense Makes Commercial Sense,' *Business Week*, February 7, 1994, p. 29. What became known in the business media as the *new* 'tech-swap' or the (Secretary of Defense) 'Perry Initiative,' where the United States accesses Japan's technology for military use has actually been going on for years.

26. The idea of convergence is discussed at length in C. Fred Bergsten and Marcus Nolan, *Reconcilable Differences: United States-Japan Economic Conflict* (Washington, D.C.: Institute for International Economics, 1993),

pp. 9-12, 210-213.

27. See Igarashi Takeshi, 'Circumventing Japan-U.S. Conflict,' *Japan Quarterly*, January-March 1991, p. 15.

28. *Japan Economic Almanac 1991* (Tokyo: *Nihon Keizai Shimbun*, 1991), p. 32.

29. Examples of works by leading U.S. revisionists are Chalmers Johnson, *MITI and the Japanese Miracle* (Stanford, CA: Stanford University Press, 1982); Clyde Prestowitz, *Trading Places* (New York: Basic Books, 1988); James Fallows, 'Containing Japan,' *The Atlantic Monthly*, May, 1989, pp. 40-54; and James Fallows, *Looking at the Sun: The Rise of the New East Asian Economic and Political System* (New York: Vintage Books, 1995).

30. Shintaro Ishihara, *The Japan That Can Say No* (New York: Simon & Schuster, 1991).

31. See 'Japan Aid Ousted; He'd Criticized Arms Role,' *The New York Times*, December 3, 1993, p. A9; and 'Defense Chief Exit Exacerbates Coalition Rift,' *The Nikkei Weekly*, December 6, 1993, p. 1.

32. See *Japan Economic Yearbook 1990* (Tokyo: *Nihon Keizai Shimbun*, 1990), p. 5.

33. See, for example, Nagayo Homma, 'The American Revisionists' Hostile View of Japan,' *Economic Eye: A Quarterly Digest of Views from Japan*, Winter 1989, pp. 19-21.

34. See James Fallows, 'Containing Japan,' *The Atlantic Monthly*, May 1989, pp. 40-54; and 'Executive Summary,' in *Harness the Rising Sun: An American Strategy for Managing Japan's Rise as a Global Power*, eds. Joseph Nye and David Rowe (Lanham, MD: University Press of America, 1993) pp. 1-21.

35. *The Japan Economic Almanac 1990* (Tokyo: *Nihon Keizai Shimbun*, 1990), pp. 36-37.

36. Subcommittee on Unfair Trade Policies Measures (SUTPM), *Report on Unfair Trade Policies by Major Trading Partners* (Tokyo: Ministry of International Trade and Industry [MITI], June 8, 1992), p. 97.

37. Ibid., p. 98; SUTPM, *Report on Unfair Trade Practices by Major Trading Partners* (Tokyo: MITI, May 11, 1993), pp. 5, 125, 200; and SUTPM, *1994 Report on Unfair Trade Policies by Major Trading Partners* (Tokyo: MITI, 1994). The most recent report is SUTPM, *1996 Report on the WTO Consistency of Trade Policies by Major Trading Partners* (Tokyo: MITI, 1996).

38. *Japan Economic Almanac 1993* (Tokyo: *Nihon Keizai Shimbun*, 1993), p. 26.

39. United Nations Industrial Development Organization, *Industry and Development, Global Report 1993/94*, p. 10.

40. See SUTPM, *Report on Unfair Trade Policies by Major Trading Partners* (1993), p. 135; and SUTPM, *1996 Report on the WTO Consistency of Trade Policies by Major Trading Partners*, pp. 114-115.

41. See 'Neither Side Budges in Framework Talks,' *The Nikkei Weekly*, January 17, 1994, p. 2. Responding to comments by former U.S. trade representative, Mickey Kantor, and how they would impact on trade discussions, a senior MITI official stated [that] 'they won't affect us at all. It is up to the U.S. whether it triggers a unilateral sanction and violates multinational rules.'

42. Laura D'Andrea Tyson, *Who's Bashing Whom? Trade Conflict in High Technology Industries* (Washington, D.C.: Institute for International Economics, 1992), pp. 55-58, 83, 255-267.

43. Quoted in 'New Japan-U.S. Era: Candor, Clarity,' *The Nikkei Weekly*, February 14, 1994, pp. 1, 19; and MITI's view on the 1996 bilateral chip dispute is stated in *Our Position on the Japan-U.S. Semiconductor Arrangement* (Tokyo: MITI, December 1995), accessed on the World Wide Web.

44. MITI, *Japan's Present Policies Regarding the Domestic and International Economy*, Address by Hiroshi Kumagai, Minister for International Trade and Industry, September 17, 1993 (to the Foreign Correspondents' Club of Japan).

45. See descriptions in 'Japan Warns U.S. Against Imposing Trade Sanctions,' *The New York Times*, February 13, 1994, pp. 1, 10; 'Japan-U.S. Relations Face Strains,' *The Daily Yomiuri*, February 13, 1994, p. 2; 'Japan to Brace for U.S. Retaliation,' *The Daily Yomiuri*, February 13, 1994, p. 2; 'Premier Says Trade Talks in the Rough,' *The Daily Yomiuri*, February 14, 1994, p. 1; and 'U.S. Strategy for Japan Talks a Failure,' *The Daily Yomiuri*, February 14, 1994, p. 2.

46. On the views of well known Japanese economists on the February 1994 trade impasse see 'Japan Should Never Give in to U.S. Demand for Targets,' *The Nikkei Weekly*, February 7, 1994, pp. 6-7.

47. See 'Emergence of Neo-Asianism Reflects Japan's Identity Search,' *The Nikkei Weekly*, January 17, 1994, pp. 1, 22; and 'Trade Surplus with Asia Tops Imbalance with the U.S.,' *The Nikkei Weekly*, April 25, 1994, p. 1.

48. Ibid.

49. See MITI, 'Policy Statement on the Openness of the Japanese Foreign

Investment Policy' (document), in *Measures for Promoting Foreign Direct Investments in Japan*, January 1992.

50. A similar point is made in Harrison Holland, *Japan Challenges America: Managing an Alliance in Crisis* (Boulder, CO: Westview Press, 1992), p. 4.

51. See *Peacekeeping and International Relations*, September/October 1993, p. 7.

52. See 'Party Attitudes Toward Major Issues' (Table), *The Nikkei Weekly*, April 18, 1994, p. 4.

53. Yasuhiro Ueki, 'Japan's UN Diplomacy: Sources of Passivism and Activism,' in *Japan's Foreign Policy After the Cold War, Coping with Change,* ed. Gerald Curtis (Armonk, N.Y.: M.E. Sharpe, 1993), pp. 358-362.

54. Ministry of Foreign Affairs, *Japan's Diplomatic Bluebook 1991: Japan's Diplomatic Activities* (Tokyo: Ministry of Foreign Affairs, 1991), pp. 31-32.

55. See 'Party Attitudes Toward Major Issues' (Table), *The Nikkei Weekly*, April 18, 1994, p. 4; and *Peacekeeping and International Relations*, January/February 1993, p. 9.

56. See 'Budget Hike for Military Smallest in 34 Years,' *The Nikkei Weekly*, February 14, 1994, p. 4; 'Defense Budget Rise Lowest in 34 Years,' *The Japan Times Weekly International Edition*, February 21-27, 1994, p. 3; 'Defense Budget Increase Held to Record Low of 0.9%,' *The Nikkei Weekly*, February 21, 1994, p. 4; and 'Socialist Make Their Mark With Low Defense Budget,' *The Nikkei Weekly*, August 1, 1994, pp. 1, 19.

57. Survey results can be found in Sasaki Yoshitaka, 'Japan's Undue International Contribution,' *Japan Quarterly*, July-September 1993, pp. 259-265; and *The Japan Economic Almanac 1994* (Tokyo: *Nihon Keizai Shimbun*), p. 20.

58. See Charles Overby, 'A Quest for Peace with Article 9,' *Japan Quarterly*, April-June 1994, p. 154.

59. See '44% in Favor of Revising the Constitution,' *The Daily Yomiuri*, March 31, 1994, p. 1.

60. For discussions on the different views of those supporting this alternative see ibid.; Koya Azumi, 'The Future of U.S.-Japan Relations: A Peaceful Alternative,' *Business and the Contemporary World*, Winter 1992, pp. 81-85; and Kitaoka, 'Opting for a Global Alliance,' *Japan Echo*, pp. 26-36.

61. See, for example, Jonathan Howe, 'Will the United States Lead a New World Order?' *The Fletcher Forum of World Affairs*, Winter/Spring 1994,

pp. 23-29; Gregory Harper, 'Creating a U.N. Peace Force: A Case for U.S. Leadership,' *The Fletcher Forum of World Affairs*, Winter/Spring 1994, pp. 49-63; and William Durch, 'Getting Involved: The Political-Military Context,' in *The Evolution of UN Peacekeeping: Case Studies and Comparative Analysis*, ed. William Durch (New York: St. Martin's 1993), pp. 16-38.

62. Office of Management and Budget, *Budget of the United States Government, Fiscal Year 1995* (Washington, D.C.: USGPO, 1994), p. 9.

63. Prime Minister's Office, *Public Opinion on Japan's Peace and Security* (Foreign Press Center, Japan, September 1989).

64. Ibid.

65. See 'U.S. Still Most Trustworthy, But Slipping,' *The Daily Yomiuri*, May 3, 1994, p. 3.

66. Ministry of Foreign Affairs, *Diplomatic Bluebook 1993: Striving for a More Secure and Humane World* (Tokyo: Ministry of Foreign Affairs, 1994), pp. 135-136. (Also accessible on the World Wide Web.)

67. Alan Tonelson, 'Superpower Without a Sword,' *Foreign Affairs*, Summer 1993, p. 177.

68. Ueki, 'Japan's Diplomacy: Sources of Passivism and Activism,' pp. 352-355.

69. See, for example, United Nations Association of the USA and the Asia Pacific Association of Japan, *The Japanese-American Alliance: A Framework for the Future* (New York: United Nations Association of the United States of America, March 1983), p. 21.

70. United Nations (U.N.), *United Nations Disarmament Yearbook*, vol. 4, 1979 (New York: U.N. Publication, 1980), p. 98.

71. U.N., *Disarmament Yearbook*, vol. 16, 1991 (New York: U.N. Publications, 1992), 118-121.

72. See U.N., *Disarmament Yearbook*, vols. 9-13, 1984-1988 (New York: U.N. Publication, various years).

73. U.N., *Disarmament Yearbook*, vol. 17, 1992 (New York: U.N. Publication, 1993), 142-143.

74. See Jorge Castaneda, trans. by Alba Amoia, *Legal Effects of United Nations Resolutions* (New York: Columbia University Press, 1969).

75. See U.N., *Disarmament Yearbook*, vol. 16, 1991, p. 490; U.N., *Disarmament Yearbook*, vol. 17, 1992, pp. 374-375.

76. U.N., *Disarmament Yearbook*, vol. 16, 1991, pp. 350-351.

77. *Japan Economic Almanac 1994* (Tokyo: *Nihon Keizai Shimbun*, 1994), p. 18.

78. U.N., *Disarmament Yearbook*, vol. 9, 1984 (New York: U.N. Publication, 1985), pp. 207-212.

79. See ibid.; and U.N., *Disarmament Yearbook*, vol. 6, 1981 (New York: U.N. Publication, 1982), p. 128.

80. Ministry of Foreign Affairs, *Diplomatic Bluebook 1991: Japan's Diplomatic Activities* (Tokyo: Ministry of Foreign Affairs, 1991), p 32.

81. *Japan Economic Almanac 1994*, p. 54.

82. Ministry of Foreign Affairs, *United Nations and Japan 1995*, accessed on the World Wide Web.

83. See 'Rethinking of Security Strategy Hits Snag,' *The Nikkei Weekly*, December 6, 1993, p. 4. However, shortly after socialist prime minister Tomiichi Murayama assumed office in June 1994, the socialist party reversed its position on the unconstitutionally of the Self Defense Forces.

84. 'Military Brass Braces for Change as Panel Begins Defense Review,' *The Nikkei Weekly*, March 7, 1994, p. 4.

85. Ibid.; 'Defense Council Faces Tough Task,' *The Daily Yomiuri*, March 2, 1994, p. 3; and 'Defense Officials Worried by Hosokawa Panel's Plans,' *The Daily Yomiuri*, March 26, 1994, p. 3.

86. See Ministry of Foreign Affairs, *United Nations and Japan 1995*, accessed on the World Wide Web; and *Comment by the Chief Cabinet Secretary on Nuclear Testing by France,* accessed on the World Wide Web. For more on Japan's position in the areas of disarmament, collective security, and its trade differences with the United States see Anthony DiFilippo, 'Why Japan Should Engage the United States: Establishing an Equal Partnership in the New Global System,' *Current Politics and Economics of Japan*, vol. 4, no. 4, 1995, pp. 223-240; Anthony DiFilippo, 'The Challenge of a Superpower: Japan's Trade Dispute with the U.S. and Its International Identity at a Crossroad, *By the Way: Bilingual Journal of U.S.-Japan Relations*, vol. 6, no. 1, December/January 1996, pp. 42-47; and Anthony DiFilippo, 'Creating Global Security: Japan as a Potential Catalyst,' in *Economics of Conflict and Peace*, eds. Jurgen Brauer and William Gissy (Aldershot, Great Britain: Avebury, forthcoming).

87. See, for example, 'Socialist Cadres Expected to Back Boss in Sweeping Reversals of Policy,' *The Nikkei Weekly*, August 1, 1994, p. 4; and Overby, 'A Quest for Peace with Article 9,' *Japan Quarterly*, pp. 149-150.

88. Ueki, 'Japan's Diplomacy: Sources of Passivism and Activism,' pp. 347-350.

89. See, for example, Ministry of Foreign Affairs, *Diplomatic Bluebook*

1992, pp, 442-445.

90. The U.N. Charter, Chapter 1, Article 1, states that the purpose of United Nations is to establish peace through the world and to do this it needs to 'take effective collective measures for the prevention and removal of threats to the peace...' See Edmund Osmanczyk, *Encyclopedia of the United Nations and International Relations* (New York: Taylor and Francis, 1990), p. 946.

91. United States Arms Control and Disarmament Agency, *Documents on Disarmament 1969* (Washington, D.C.: USGPO, 1970), p. 308.

92. 'New U.S. Peacekeeping Policy De-emphasizes Role of the U.N.,' *The New York Times*, May 6, 1994, pp. A1, A7; for proposed (republican) restrictions that would further retard the U.N.'s effectiveness in international security, see 'U.N. Voices Concern Over a Bill to Limit U.S. Peacekeeping Role,' *The New York Times*, February 10, 1995, p. A9.

93. See National Science Foundation, *Tokyo Report Memorandum*, June 3, 1992.

94. Ministry of Foreign Affairs, *Diplomatic Bluebook 1992*, p. 19.

95. DiFilippo, 'Working with Japan to Revitalize Industry,' *IEEE Technology and Society Magazine*, pp. 21-27.

96. 'U.N. Asks Japan to Pay Dues Early,' *Japan Times Weekly International Edition,* May 13-19, 1996, p. 2.

97. See 'U.N. Chief Says Peacekeeping Unrelated to Permanent Seat,' *The Japan Times Weekly International Edition*, December 27, 1993 - January 2, 1994, p. 3.

98. See Seymour Melman, *Profits Without Production* (New York: Knopf, 1983).

99. See Anthony DiFilippo, 'Technology and Trade Problems in Postindustrial Economy: Japan Versus the United States,' in *Perspectives and Issues in International Political Economy*, ed. Chronis Polychroniou (Westport, CT: Praeger, 1992), pp. 150-171.

100. See *Japan Economic Almanac 1994*, p. 17.

101. See, for example, Michael Blaker, 'Evaluating Japan's Diplomatic Performance,' in *Japan's Foreign Policy: After the Cold War*, pp. 1-42.

102. See Center for Defense Information, The Defense Monitor, *Policing World Trouble Spots: United States or United Nations*? vol. XXIII, no. 3, 1994, pp. 1-2.

Index